M000158290

OUTER SPACE: 100 POEMS

✳

Poets and astronomers often ask the same questions. Where did we come from? Why are we here? Where are we going? Throughout human history, poetry has provided stories about what people observe in the sky. Stars, planets, comets, the moon, and space travel are used as metaphors for our feelings of love, loneliness, adventurousness, and awe. This anthology includes poets, astronomers, and scientists from the twelfth century BCE to today, from all around the world. Sappho, Du Fu, Hafez, and Shakespeare are joined by Gwyneth Lewis's space requiem, Tracy K. Smith on the Hubble telescope, and Charles Simic, whose poem accompanied a NASA mission. Astronomers Tycho Brahe and Edmund Halley accompany modern scientists including Rebecca Elson, Alice Gorman on the first woman in space, and Yun Wang's space journal on travel to Andromeda. This collection reaches across time and cultures to illuminate how we think about outer space, and ourselves.

MIDGE GOLDBERG has written and published three poetry collections, including, in 2021, *To Be Opened After My Death*. She was the recipient in 2016 of the Richard Wilbur Poetry Award for her book *Snowman's Code*, and, in the same year, of the Howard Nemerov Sonnet Award. Her poems have appeared in *Measure*, *Alabama Literary Review*, *Light*, *Poetry Speaks: Who I Am*, and other national journals and anthologies, as well as on Garrison Keillor's daily podcast and newsletter, *The Writer's Almanac*. She is a graduate of Yale University and in addition holds an MFA from the University of New Hampshire.

The Flammarion woodcut (or engraving) appeared in an 1888 book by French astronomer Nicolas Camille Flammarion. (Photo by Fine Art Images / Heritage Images / Getty Images)

OUTER SPACE
100
POEMS

EDITED BY

Midge Goldberg

CAMBRIDGE
UNIVERSITY PRESS

CAMBRIDGE
UNIVERSITY PRESS

Shaftesbury Road, Cambridge CB2 8EA, United Kingdom

One Liberty Plaza, 20th Floor, New York, NY 10006, USA

477 Williamstown Road, Port Melbourne, VIC 3207, Australia

314–321, 3rd Floor, Plot 3, Splendor Forum, Jasola District Centre,
New Delhi – 110025, India

103 Penang Road, #05–06/07, Visioncrest Commercial, Singapore 238467

Cambridge University Press is part of the University of Cambridge.

It furthers the University's mission by disseminating knowledge in the pursuit of
education, learning, and research at the highest international levels of excellence.

www.cambridge.org
Information on this title: www.cambridge.org/9781009203609
DOI: 10.1017/9781009203616

First published 2022

Printed in the United Kingdom by TJ Books Limited, Padstow Cornwall

A catalogue record for this publication is available from the British Library.

ISBN 978-1-009-20360-9 Hardback

To my stargazers, Hannah, Ely, and Bob

Contents

✳

Contents

Contents

Contents

Contents

Contents

Contents

Contents

INTRODUCTION

✳

Outer space has always captured my imagination. As a child I saw the first moon landing on our old black and white television. When I was in high school in Florida, our teachers let us go outside to watch the bright line that streaked up into the sky when the first space shuttle took off. I followed the discoveries of the Hubble telescope, the Mars rover, and other explorations, and have always been a fan of science fiction books and movies. When I started to write poetry, those interests found their way into my poems. I often wondered about the link between poetry and outer space, so I was delighted to have the chance to edit an anthology of space poems. While poetry and astronomy may seem unrelated, in some respects they are just different ways of responding to the same questions – where did we come from, why are we here, where are we going?

Objects in outer space have appeared in poetry since the earliest poems were written. Poetry provided stories about the inexplicable and wondrous entities and events people observed in the sky, and still tries to convey the unimaginably vast distances and sizes and the mind-bending hypotheses involved in comprehending outer space today. Conversely, stars, planets, comets, the sun, the moon, and space travel have all been used as metaphors for our emotions and questions. As humans came to understand more of what they observed, the imagery changed – or didn't, because no matter how old or how new the poems are, whether they are about Greek gods or modern technology, outer space provides metaphors that persist over time for our very human emotions of love, faith, loneliness, adventurousness, and awe.

This anthology reaches across time and cultures to try to show how we have thought about the universe and our place in it, and how we have expressed our thoughts in poetry. It is organized chronologically by poet's birthdate, in the hope of showing how the poems change, contradict each other, and evolve, but also remain the same and sometimes even talk to each other across the ages. Where the date is uncertain, I have indicated the reason for its placement. I selected poems for their craft and beauty, topics, culture of origins, and perspectives. They come in many forms, including blank verse, sonnets, metrical passages from plays, free verse, and prose poems. There are a number of poems I had to leave out that should have been included, and I'm truly sorry for the omissions. But I hope you will enjoy the selections and go investigate when a poem captures your attention.

You can, of course, read this anthology any way you want. You can look for a particular topic, or let the book fall open where it will. If you read it chronologically, you may find some surprising connections. But I would like to draw your attention to some that resonated with or unsettled me.

It should come as no surprise that this anthology contains a number of love poems. The very first poem in the book, from Egypt, around the twelfth century BCE, compares the loveliness of the beloved to Sirius, the brightest star in the sky. Venus and the moon are favorite images for their beauty and their help to lovers, though the moon also appears as a symbol for lost and changeable love from Sappho to the present. Two poems I found very moving are "Love" by Thoreau, in which a double star represents the lovers, and "Delay" by Elizabeth Jennings, a twentieth-century poet, who uses the speed of starlight to talk about the lifespan of love. Our understanding of the phenomena of outer space may have grown, but love is always love!

As you can imagine, where we came from and how this world came to be are favorite topics for poets. Poems from many times and cultures have described the creation of the moon, stars, and planets by divine powers, including Psalm 8, Milton's *Paradise Lost*, a chant from the Dinka of Nilotic Sudan, and "A Song of the Navajo Weaver" by Bertrand N. O. Walker (Hen-tah). While I have no expert knowledge of astronomy or the evolution of thought in this field, I could see as I collected these poems that, as new discoveries were made and beliefs and philosophies changed, they appeared and were debated in poems too. Lucretius, a first-century BCE Latin poet, talks of atoms spontaneously and randomly combining to become the greater bodies in outer space. Edmund Halley, the seventeenth-century astronomer, while still attributing creation to God, praised Isaac Newton for his discovery of gravity and all the resultant understanding of orbits. One of the joys of putting this anthology together came from watching the poems "talk" to each other. While I don't know if Mercy Otis Warren, an American poet born about seventy years after Halley, actually read Halley's poem, her poem, coming right after his chronologically, seems like a rebuttal, minimizing Newton in favor of God as sole creator of the universe.

Why are we here? What is our place in this universe? Looking at outer space — its vast distances, the blackness, the objects separated by incomprehensible distances — drives some poets to answer these questions; others are terrified by them. Frances Ellen Watkins Harper, a nineteenth-century African-American abolitionist, wrote with conviction about her belief that, despite our short lives as compared to the age of the stars, God created every person to have a place and be as important as any star, if not more important. However, in Edna St. Vincent Millay's "Renascence" (of which just an extract is here — I encourage you to read the whole poem), published only about seventeen years after Harper's, the speaker screams, overwhelmed at looking at the sky and seeing the infinity and eternity there.

As modern exploration reveals more and more of the immensity of the universe, poetry reflects the uncertainty and fear those discoveries brought about. Tracy K. Smith, in her poem "My God, It's Full of Stars" (again, this is only an extract and I encourage you to read the whole poem), speaks of her father's work on the Hubble telescope, and of the terror in seeing to the end of the universe, while Richard Wilbur experiences but then overcomes that fear in "In the Field." A number of poems wonder if technological advances and scientific knowledge that allow us to see more of the universe actually get in the way of perceiving its beauty and power. A favorite of mine has a different perspective — in "Carnal Knowledge," Rebecca Elson, an astronomer, removes the fear of infinity by naming its causes, but longs to viscerally feel that same thrill in the body.

Another reaction to looking at outer space is to look back at one's home. In Du Fu's eighth-century poem translated by Yun Wang, a cosmologist, and in Wang's own poem, objects in space evoke the poets' gardens, symbols for home. Rhina Espaillat talks to her long-in-the-future great-great-grandson the space pioneer, reminding him not to forget the beauty of Earth. In Allison Joseph's "My Father's Kites," Earth and space have a tricky, delicate relationship, while Claude McKay dismisses the moon as antique and overly precious compared to a modern city.

Poems throughout the anthology celebrate the idea of exploration into the unknown. It was great fun to discover in Dante's *Inferno* the phrase "the unpeopled world," spoken by Ulysses to his sailors, and then find that phrase across the centuries in the works of poets Anna Laetitia Barbauld, Howard Nemerov, and Bill Coyle. In Barbauld's poem, "A Summer Evening's Meditation," she describes leaving Earth and passing and then going beyond all the planets. Helen Hunt Jackson in the nineteenth century wrote about getting on a spaceship and actually emigrating to another star!

Once space exploration began, poets based poems on the experiences of actual astronauts and cosmonauts. Alice Gorman, a space archeologist, has an amazing poem about Valentina Tereshkova, the first woman in space, taken from the transcript of her space flight. Howard Nemerov, US Poet Laureate at the time, was commissioned by NASA to attend the 1985 launch of the space shuttle *Atlantis* and document it in a poem. Gwyneth Lewis weaves together stanzas about her cousin's experience as part of the crew of a space shuttle sent to repair the Hubble. Along with travelling in space comes imagining the aliens we might meet there — two humorous (and dark!) poems contain the perspective of those aliens, "Earth" by Oliver Herford, and "The First Men on Mercury" by Edwin Morgan.

Two contemporary poems round out the anthology by looking both back and forward. Bill Coyle's "C. E." reflects from a modern perspective on how we thought about outer space, invoking some of the earlier poems in this volume, including those by Dante, Milton, Keats, and a reference to Auden's "Moon Landing" (not included here, but I recommend you read it on your own). And Sarah Howe's poem "Relativity," commissioned for Britain's 2015 National Poetry Day, dedicated to and read by Stephen Hawking (I encourage you to find this video online), finds the poetry in what we have learned and what we still have to learn.

I'm honored (and lucky) to be able to end the volume with Charles Simic's prose poem, which was actually sent into outer space on *Lucy*, a NASA space probe, in July 2021. It combines much of what the poems in this anthology try to address — the wonder of what is out there, the love of home, and the hope that these things will be what we have in common with whoever may come across that poem thousands of years from now.

OUTER SPACE: 100 POEMS

✳

COLLECTED SONGS FROM THE BALLAD-MISTRESS, "RETICENT LOVERS"

✳

Boy

There is no **one** like my sister,
    the loveliest of all women.
She is Sirius rising
    heralding the rising Nile.
Shining, light-skinned, exquisite,
    her eyes constantly wandering.
Her lips speak sweetly
    but do not ramble.
Her neck is long, her breasts white
    her hair lapis lazuli.
Her arms outshine gold,
    with lotus-fingers,
full rear, narrow waist,
    and thighs that lead to splendor.
Her stride is a magic-charm;
    my heart is bewitched in her embrace.
She'll turn the head of any man
    who catches a glimpse of her.

Any man who gets her into bed
      is indeed a master lover.
When she emerges
      it is like the emergence of the **One**.

UNKNOWN (ANCIENT EGYPTIAN LOVE SONGS,
CHESTER BEATTY I PAPYRUS)

Translation by Alan Humm

Sirius   Sirius was an important part of the Egyptian astronomical calendar. The brightest star in the sky, its annual summer appearance indicated the imminent flooding of the Nile, crucial to the agricultural year.

2

PSALM 8: FOR THE LEADER; UPON THE GITTITH. A PSALM OF DAVID

✳

O the Lord, our Lord, how glorious is Thy name in all the earth! whose majesty is rehearsed above the heavens.

Out of the mouth of babes and sucklings hast Thou founded strength, because of Thine adversaries; that Thou mightest still the enemy and the avenger.

When I behold Thy heavens, the work of Thy fingers, the moon and the stars, which Thou hast established;

What is man, that Thou art mindful of him? and the son of man, that Thou thinkest of him?

Yet Thou hast made him but little lower than the angels, and hast crowned him with glory and honour.

Thou hast made him to have dominion over the works of Thy hands; Thou hast put all things under His feet:

Sheep and oxen, all of them, yea, and the beasts of the field;

The fowl of the air, and the fish of the sea; whatsoever passeth through the paths of the seas.

O the Lord, our Lord, how glorious is Thy name in all the earth!

Translation by Jewish Publication Societies

3

FROM *THE ILIAD*, BOOK XVIII

✳

For here he placed the earth and heaven, and here
The great deep and the never-resting sun
And the full moon, and here he set the stars
That shine in the round heaven—the Pleiades,
The Hyades, Orion in his strength,
And the Bear near him, called by some the Wain,
That, wheeling, keeps Orion still in sight,
Yet bathes not in the waters of the sea.

HOMER

Translation by William Cullen Bryant

4

SAPPHO 34

✳

See the stars that gather around the young moon
shrink and hide their own little sparks of brightness
when the moon now grown to its greatest fullness
turns earth to silver.

SAPPHO

Translation by Michael Ferber

5

FROM *DE RERUM NATURA*, BOOK 2

✳

In no way should it be thought likely
(since infinite space lies empty in all directions,
and in its boundless depths atoms in innumerable number
fly about in many ways, propelled in eternal motion)
that this earth and sky were created alone,
and those elements of matter in outer space do nothing;
various atoms, in violently colliding—spontaneous, aimless,
randomly driven—to no purpose,
which eventually combine unexpectedly, and
hurled together, are forever becoming the source of greater bodies
of earth, sea, sky and the beginnings of animals.

LUCRETIUS

Translation by Deborah Warren

6

FROM *CARMINA*, 66

✳

That man who first knew all the stars that light the skies;
    Who found out when they set and when they rise,
And how the all-consuming sun's bright flames go dark,
    And why stars fade when they have passed some mark;
How Dian leaves her course when dulcet love has called
    (She's off on Latmos, secretly installed):
That very Conon found me out in heaven's floor—
    That lock of Berenice's that she tore,
And, shining brightly, that she'd promised all the gods,
    Pleading with silken arms for heaven's nods,
Back when the king, blessed in his nuptial joy, had told
    Of how he'd come down on the Assyrian fold
Still bearing traces of the lovely nightly broils
    He'd ventured on to win his virgin spoils.

CATULLUS

Translation by Len Krisak

That very Conon found me … The speaker of this poem is the lock of hair from the
constellation *Coma Berenices*, Berenice's hair, named by Conon of Samos, a Greek astronomer.

7

FROM *GEORGICS*, BOOK 4

✷

Two seasons for harvesting the fertile crops:
When the Pleiad Taygete, showing to the earth
her sweet face, has repelled with scornful feet
the Ocean tides; and when, fleeing watery Pisces,
she sinks from the sky into the wintry waves.

VIRGIL

Translation by Deborah Warren

8

✳

A cat laps
at beams in a bowl.
An elephant takes
rays amid leaves to be
sweet white stems. A girl
resting from love, reaches
for the bright pool of her
discarded dress.
So by moonlight are we all,
greater or less,
deceived.

[BHĀSA?]

Translation from the Sanskrit by Victoria Moul

9

FULL MOON

✳

A lone moon full over the balcony
reflected by a cold river onto a night window
Undulating waves shift gold
Intricate shimmers transform the lit quilt
The flawless moon in empty mountain quiet
hangs high against dim constellations
The old garden's pines and osmanthus bloom
ten thousand miles away in the same crystal light

DU FU

Translation by Yun Wang

FROM "THE OLD ENGLISH PHOENIX"

✳

Those woods are watched   by a wondrous, delightful,
strong-winged fowl   called a "phoenix."
He lives in that place   alone in a lair
in steadfast style.   He will know no death
on that pleasant plain   while the Earth endures.
There he must scan   the cycle of the sun
and keep on coming   toward God's candle,
that joyous gem,   to observe zealously—
up until the rise   of the most regal star
above the brilliant   eastern sea-waves—
our Father's legacy   gleaming gorgeously
as God's bright sign.   The star is hidden,
lurking below   the waves in the west.
It stays through sunrise,   and the black night
glides into gloom,   then the stately bird,
strong of wing,   searches below the sky
and over ocean   for mountain streams
until the heavens' rays   arise from the east
gliding upon   open ocean,

UNKNOWN

Translation from the Old English by A. M. Juster

FROM *INFERNO*: CANTO XXVI

✶

'O brothers, who amid a hundred thousand
    Perils,' I said, 'have come unto the West,
    To this so inconsiderable vigil
Which is remaining of your senses still,
    Be ye unwilling to deny the knowledge,
    Following the sun, of the unpeopled world.
Consider ye the seed from which ye sprang;
    Ye were not made to live like unto brutes,
    But for pursuit of virtue and of knowledge.'
So eager did I render my companions,
    With this brief exhortation, for the voyage,
    That then I hardly could have held them back.
And having turned our stern unto the morning,
    We of the oars made wings for our mad flight,
    Evermore gaining on the larboard side.
Already all the stars of the other pole
    The night beheld, and ours so very low
    It did not rise above the ocean floor.
Five times rekindled and as many quenched
    Had been the splendor underneath the moon,
    Since we had entered into the deep pass,
When there appeared to us a mountain, dim
    From distance, and it seemed to me so high
    As I had never any one beheld.
Joyful were we, and soon it turned to weeping;
    For out of the new land a whirlwind rose,
    And smote upon the fore part of the ship.

Three times it made it whirl with all the waters,
    At the fourth time it made the stern uplift,
    And the prow downward go, as pleased Another,
Until the sea above us closed again.

DANTE

Translation by Henry Wadsworth Longfellow

12

I SAW THE GREEN FIELDS OF THE SKY ...

✳

I saw the green fields of the sky,
           and there a sickle moon –
I reckoned what I'd sown, and thought,
           "The harvest will come soon."

I said, "My luck, you've been asleep,
           now dawn has brought the sun."
She said, "The past is past; do not
           despair of all you've done;

The night you leave this world, ascend
           like Jesus through the skies –
Your lamp, a hundred times, will light
           the sun as you arise.

Don't trust the morning star, she is
           a highway robber who
Stole Kay Kavus's throne, and then
           the belt of Khosrow too.

Gold earrings set with rubies may
           charm you, and lead you on,
But know this: Beauty's reign is brief,
           and all too quickly gone."

God keep the evil eye from your
           sweet beauty, which can field
A pawn to make the sun and moon
           precipitously yield.

Say to the heavens, "Don't boast of splendor!"
          When love is matched with you,
The harvest of the moon's a grain,
          and of the stars but two.

Hypocrisy will burn the harvest
          religion reaped; and so,
Hafez, shrug off this Sufi cloak —
          just leave now, let it go.

HAFEZ

Translation by Dick Davis

FROM *THE CANTERBURY TALES,* "THE NUN'S
PRIEST'S TALE"

✳

Whan that the month in which the world bigan,
That highte March, whan god first maked man,
Was complet, and [y]-passed were also,
Sin March bigan, thritty dayes and two,
Bifel that Chauntecleer, in al his pryde,
His seven wyves walking by his syde,
Caste up his eyen to the brighte sonne,
That in the signe of Taurus hadde y-ronne
Twenty degrees and oon, and somwhat more;
And knew by kynde, and by noon other lore,
That it was pryme, and crew with blisful stevene.
"The sonne," he sayde, "is clomben up on hevene
Fourty degrees and oon, and more, y-wis.

GEOFFREY CHAUCER

14

THOSE WHO BUILD HOUSES AND TEMPLES

✳

Essential
to count the haab years or katuns
that have passed since
the great powerful men
raised the walls of the ancient cities
that we see now
here in the Province of the Plains,
all these cities scattered
on the earth
here and there, on high hills.

Here in the cities, we try to give
meaning to what we see today in the skies
and what we know;
for day to day
at midday
we see in the skies
the signs told to us by
the ancient people of this land,
the ancient people of these villages
here on our earth.

Let us purify our hearts
so at nightfall,
and at midnight,
from horizon to zenith
we may read the face of the sky.

UNKNOWN

Translation from the Mayan by John Curl

It is estimated this poem was written in the 1400s. The only specific date in the manuscript is 1440, when the ancestor of the collector of these songs and poems was the assistant village chief, and the collector says they come from that period.

15

LOVE THE LIGHT GIVER

✳

With your fair eyes a charming light I see,
     For which my own blind eyes would peer in vain;
     Stayed by your feet the burden I sustain
     Which my lame feet find all too strong for me;
Wingless on your pinions forth I fly;
     Heavenward your spirit stirreth me to strain;
     E'en as you will I blush and blanch again,
     Freeze in the sun, burn 'neath a frosty sky.
Your will includes and is the lord of mine;
     Life to my thoughts within your heart is given;
     My words begin to breathe upon your breath;
Like to the moon am I, that cannot shine
     Alone; for lo! your eyes see nought in heaven
     Save what the living sun illumineth.

MICHELANGELO

Translation by John Addington Symonds

16

URANIA

✳

Urania, seeing this cave from heaven, asks:
What new trick is being prepared on earth?
And, gliding down, says: Why hide heaven's stars
when the earth steals into my sacred home?
If Heaven lays bare what earth hides in its heart,
Why hide the sky's stars in such curved and obscure orbits?—
When earth's low depths see into my lofty kingdom,
when even the Earth gives a path to the highest stars?

TYCHO BRAHE

Translation by Deborah Warren

FROM "EPITHALAMION"

✳

Ah when will this long weary day have end,
And lende me leave to come unto my love?
How slowly do the houres theyr numbers spend?
How slowly does sad Time his feathers move?
Hast thee O fayrest Planet, to thy home
Within the Westerne fome:
Thy tyred steedes long since have need of rest.
Long though it be, at last I see it gloome,
And the bright evening star with golden creast
Appeare out of the East.
Fayre childe of beauty, glorious lampe of love
That all the host of heaven in rankes doost lead,
And guydest lovers through the nights sad dread,
How chearefully thou lookest from above,
And seemst to laugh atweene thy twinkling light
As joying in the sight
Of these glad many which for joy doe sing,
That all the woods them answer and their echo ring!

EDMUND SPENSER

18

ASTROPHIL AND STELLA: 31 WITH HOW SAD
STEPS, O MOON, THOU CLIMB'ST THE SKIES

✳

With how sad steps, O Moon, thou climb'st the skies!
  How silently, and with how wan a face!
  What, may it be that even in heav'nly place
That busy archer his sharp arrows tries?
Sure, if that long-with-love-acquainted eyes
  Can judge of love, thou feel'st a lover's case;
  I read it in thy looks; thy languished grace
To me, that feel the like, thy state descries.
Then, even of fellowship, O Moon, tell me,
  Is constant love deemed there but want of wit?
Are beauties there as proud as here they be?
  Do they above love to be loved, and yet
Those lovers scorn whom that love doth possess?
Do they call virtue there ungratefulness?

SIR PHILIP SIDNEY

FROM *THE TRAGICAL HISTORY OF DOCTOR FAUSTUS*, SCENE 6

✳

FAUSTUS.

…

Come, Mephistophilis, let us dispute again,
And reason of divine astrology.
Speak, are there many spheres above the moon?
Are all celestial bodies but one globe,
As is the substance of this centric earth?

MEPHIST. As are the elements, such are the heavens,
Even from the moon unto th' empyreal orb,
Mutually folded in each other's spheres,
And jointly move upon one axletree,
Whose termine is term'd the world's wide pole;
Nor are the names of Saturn, Mars, or Jupiter
Feign'd, but are erring stars.

FAUSTUS. But have they all one motion, both situ et tempore?

MEPHIST. All move from east to west in four-and-twenty
hours upon the poles of the world; but differ in their motions
upon the poles of the zodiac.

FAUSTUS. These slender questions Wagner can decide:
Hath Mephistophilis no greater skill?
Who knows not the double motion of the planets?
That the first is finish'd in a natural day;
The second thus; Saturn in thirty years; Jupiter in twelve;
Mars in four; the Sun, Venus, and Mercury in a year; the Moon

in twenty-eight days. These are freshmen's questions. But
tell me, hath every sphere a dominion or intelligentia?

MEPHIST. Ay.

FAUSTUS. How many heavens or spheres are there?

MEPHIST. Nine; the seven planets, the firmament, and the empyreal
heaven.

FAUSTUS. But is there not coelum igneum et crystallinum?

MEPHIST. No, Faustus, they be but fables.

FAUSTUS. Resolve me, then, in this one question; why are not
conjunctions, oppositions, aspects, eclipses, all at one time,
but in some years we have more, in some less?

MEPHIST. Per inoequalem motum respectu totius.

FAUSTUS. Well, I am answered. Now tell me who made the world?

MEPHIST. I will not.

FAUSTUS. Sweet Mephistophilis, tell me.

MEPHIST. Move me not, Faustus.

FAUSTUS. Villain, have I not bound thee to tell me any thing?

MEPHIST. Ay, that is not against our kingdom; this is.
Thou art damned; think thou of hell.

FAUSTUS. Think, Faustus, upon God that made the world.

MEPHIST. Remember this.
   [Exit.]

CHRISTOPHER MARLOWE

FROM *JULIUS CAESAR*,
ACT 3, SCENE 1

✳

CAESAR.

. . .

But I am constant as the northern star,
Of whose true-fix'd and resting quality
There is no fellow in the firmament.
The skies are painted with unnumber'd sparks.
They are all fire and every one doth shine,
But there's but one in all doth hold his place.

WILLIAM SHAKESPEARE

FROM "THE ANATOMY OF THE WORLD"

✳

For, before God had made up all the rest,
Corruption ent'red, and deprav'd the best.
It seized the angels, and then first of all
The world did in her cradle take a fall,
And turn'd her brains, and took a general maim,
Wronging each joint of th' universal frame.
The noblest part, man, felt it first; and then
Both beasts and plants, cursed in the curse of man.
So did the world from the first hour decay;
That evening was beginning of the day.
And now the springs and summers which we see,
Like sons of women after fifty be.
And new philosophy calls all in doubt;
The element of fire is quite put out;
The sun is lost, and th' earth, and no man's wit
Can well direct him where to look for it.
And freely men confess that this world's spent,
When in the planets and the firmament
They seek so many new; they see that this
Is crumbled out again to his atomies.
'Tis all in pieces, all coherence gone,
All just supply, and all relation;
Prince, subject, father, son, are things forgot,

For every man alone thinks he hath got
To be a phoenix, and that then can be
None of that kind, of which he is, but he.
This is the world's condition now...

JOHN DONNE

CYNTHIA'S REVELS: QUEEN AND HUNTRESS, CHASTE AND FAIR

✳

Queen and huntress, chaste and fair,
Now the sun is laid to sleep,
Seated in thy silver chair
State in wonted manner keep:
    Hesperus entreats thy light,
    Goddess excellently bright.

Earth, let not thy envious shade
Dare itself to interpose;
Cynthia's shining orb was made
Heaven to clear when day did close:
    Bless us then with wished sight,
    Goddess excellently bright.

Lay thy bow of pearl apart
And thy crystal-shining quiver;
Give unto the flying hart
Space to breathe, how short soever:
    Thou that mak'st a day of night,
    Goddess excellently bright.

BEN JONSON

Cynthia Goddess of the moon.

FROM *PARADISE LOST*, BOOK X

✳

While the Creator calling forth by name
His mightie Angels gave them several charge,
As sorted best with present things. The Sun
Had first his precept so to move, so shine,
As might affect the Earth with cold and heat
Scarce tollerable, and from the North to call
Decrepit Winter, from the South to bring
Solstitial summers heat. To the blanc Moone
Her office they prescrib'd, to th' other five
Thir planetarie motions and aspects
In *Sextile*, *Square*, and *Trine*, and *Opposite*,
Of noxious efficacie, and when to joyne
In Synod unbenigne, and taught the fixt
Thir influence malignant when to showre

JOHN MILTON

FROM *THE SONG OF THE THREE CHILDREN,*
PARAPHRAS'D

*

Ye glitt'ring Stars, who float in liquid Air,
Both ye that round the Sun in diff'rent Circles move,
   And ye that shine like Suns above;
Whose Light and Heat attending Planets share:
In your high Stations your Creator praise,
   While we admire both him and you;
Tho' vastly distant, yet our Eyes we raise,
   And wou'd your lofty Regions view;
Those immense Spaces which no Limits know,
Where purest *Æther* unconfin'd doth flow;
But our weak Sight cannot such Journies go:
'Tis Thought alone the Distance must explore;
Nothing but That to such a Height can soar,
Nothing but That can thither wing its Way,
   And there with boundless Freedom stray,
And at one View Ten thousand sparkling Orbs survey,
Innumerable Worlds and dazling Springs of Light.
O the vast Prospect! O the charming Sight!
   How full of Wonder, and Delight!
How mean, how little, does our Globe appear!

LADY MARY CHUDLEIGH

ON THE INCOMPARABLE ISAAC NEWTON

✳

his brilliant work in mathematics and physics the glory of our age and
people.

The law of the Pole, the weights of heaven's mass,
Calculated by you, formed by Jove; laws which God,
Creator, source of primordial things, has fixed:
the foundations of his eternal handiwork.
You open the inner secrets of the sky,
No force that turns the farthest orbs lies hidden.
The sun on his throne draws all of these down toward him,
pulling their starry courses from their straight paths,
and allows them to move through the vast emptinesses;
but draws to his center in unmoving circles.
Now the curved paths of the once-feared comets
Are known to us, and we are no longer awed
by the phenomena of the bearded stars.
At last we learn why the cause of silver moon's
uneven steps; why until now she refused
to submit harnessed by numbers to an astronomer:
why through the seasons stars leave and return.
We learn how Cynthia's changes pull the tides,
the flux of the sea, when the breaking waves reveal
the seaweed, bare the sand watched for by sailors,
and drive the seas in cycles far up the shore.

EDMUND HALLEY

Translation by Deborah Warren

ON A SURVEY OF THE HEAVENS

✳

Does there an infidel exist?
Let him look up—he can't resist,
These proofs of Deity—so clear,
He must the architect revere,
Whene'er to heaven he lifts his eyes,
And there surveys the spangled skies;
The glitt'ring stars, the worlds that shine,
And speak their origin divine,
Bid him adore, and prostrate fall,
And own one Lord, supreme o'er all.

One God this mighty fabrick guides,
Th' etherial circles he divides;
And measures out the distant bound,
Of each revolving planet's round;
Prevents the universal jar,
That might from one eccentric star,
Toss'd in the wide extended space,
At once—a thousand worlds displace.

What else supports the rolling spheres;
Nought but Almighty power appears,
The vast unnumber'd orbs to place,
And scatter o'er the boundless space,
Myriads of worlds of purer light,
Our adoration to excite,
And lead the wandering mind of man,
To contemplate the glorious plan.

Not even Newton's godlike mind,
Nor all the sages of mankind,
Could e'er assign another cause,
Though much they talk of nature's laws;
Of gravity's attractive force,
They own the grand, eternal source,
Who, from the depths of chaos' womb,
Prepar'd the vaulted, spacious dome;
He spake—a vast foundation's laid,
And countless globes thereon display'd.

His active power still sustains
Their weight, amidst the heavenly plains;
Infinite goodness yet protects,
All perfect wisdom still directs
Their revolutions;—knows the hour,
When rapid time's resistless pow'r,
In mighty ruin will involve,
And God—this grand machine dissolve.

Then time and death shall both expire,
And in the universal fire,
These elements shall melt away,
To usher in eternal day.

Amazing thought!—Is it decreed,
New earth and heavens, shall these succeed?
More glorious far—still more august!
In his omnific arm we trust.

But how this system 'twill excel,
Nor Angel's voice, or tongue can tell;
Nor human thought so high can soar;
His works survey, and God adore.

MERCY OTIS WARREN

27

FROM "A SUMMER EVENING'S MEDITATION"

*

Seiz'd in thought,
On fancy's wild and roving wing I sail,
From the green borders of the peopled earth,
And the pale moon, her duteous fair attendant;
From solitary Mars; from the vast orb
Of Jupiter, whose huge gigantic bulk
Dances in ether like the lightest leaf;
To the dim verge, the suburbs of the system,
Where cheerless Saturn 'midst his wat'ry moons
Girt with a lucid zone, in gloomy pomp,

Sits like an exil'd monarch: fearless thence
I launch into the trackless deeps of space,
Where, burning round, ten thousand suns appear,
Of elder beam; which ask no leave to shine
Of our terrestrial star, nor borrow light
From the proud regent of our scanty day;
Sons of the morning, first-born of creation,
And only less than HIM who marks their track,
And guides their fiery wheels. Here must I stop,
Or is there aught beyond? What hand unseen
Impels me onward thro' the glowing orbs
Of habitable nature, far remote,
To the dread confines of eternal night,
To solitudes of vast unpeopled space,
The desarts of creation, wide and wild;
Where embryo systems and unkindled suns
Sleep in the tomb of chaos? fancy droops,

And thought astonish'd stops her bold career.
But oh thou mighty mind! whose powerful word
Said, thus let all things be, and thus they were,
Where shall I seek thy presence? how unblam'd
Invoke thy dread perfection?
Have the broad eye-lids of the morn beheld thee?
Or does the beamy shoulder of Orion
Support thy throne? O look with pity down
On erring guilty man; not in thy names
Of terror clad; not with those thunders arm'd
That conscious Sinai felt, when fear appall'd
The scattered tribes; thou hast a gentler voice,
That whispers comfort to the swelling heart,
Abash'd, yet longing to behold her Maker.

   But now my soul unus'd to stretch her powers
In flight so daring, drops her weary wing,
And seeks again the known accustom'd spot,

Drest up with sun, and shade, and lawns, and streams,
A mansion-fair and spacious for its guest,
And full replete with wonders. Let me here
Content and grateful, wait th' appointed time
And ripen for the skies: the hour will come
When all these splendours bursting on my sight
Shall stand unveil'd, and to my ravish'd sense
Unlock the glories of the world unknown.

ANNA LAETITIA BARBAULD

28

THREATENING SIGNS

✳

If Venus in the evening sky
Is seen in radiant majesty,
If rod-like comets, red as blood,
Are 'mongst the constellations view'd,
Out springs the Ignoramus, yelling:
"The star's exactly o'er my dwelling!
What woeful prospect, ah, for me!"
Then calls his neighbour mournfully:
"Behold that awful sign of evil,
Portending woe to me, poor devil!
My mother's asthma ne'er will leave her,
My child is sick with wind and fever;
I dread the illness of my wife,
A week has pass'd, devoid of strife,—
And other things have reach'd my ear;
The Judgment Day has come, I fear!"

His neighbour answered: "Friend, you're right!
Matters look very bad to-night.
Let's go a street or two, though, hence,
And gaze upon the stars from thence."—
No change appears in either case.
Let each remain then in his place,
And wisely do the best he can,
Patient as any other man.

JOHANN WOLFGANG VON GOETHE

Translation by Hjalmar Boyesen

29

ON IMAGINATION

＊

Thy various works, imperial queen, we see,
   How bright their forms! how deck'd with pomp by thee!
Thy wond'rous acts in beauteous order stand,
And all attest how potent is thine hand.

   From *Helicon's* refulgent heights attend,
Ye sacred choir, and my attempts befriend:
To tell her glories with a faithful tongue,
Ye blooming graces, triumph in my song.

   Now here, now there, the roving *Fancy* flies,
Till some lov'd object strikes her wand'ring eyes,
Whose silken fetters all the senses bind,
And soft captivity involves the mind.

   Imagination! who can sing thy force?
Or who describe the swiftness of thy course?
Soaring through air to find the bright abode,
Th' empyreal palace of the thund'ring God,
We on thy pinions can surpass the wind,
And leave the rolling universe behind:
From star to star the mental optics rove,
Measure the skies, and range the realms above.
There in one view we grasp the mighty whole,
Or with new worlds amaze th' unbounded soul.

   Though *Winter* frowns to *Fancy's* raptur'd eyes
The fields may flourish, and gay scenes arise;
The frozen deeps may break their iron bands,
And bid their waters murmur o'er the sands.

Fair *Flora* may resume her fragrant reign,
And with her flow'ry riches deck the plain;
Sylvanus may diffuse his honours round,
And all the forest may with leaves be crown'd:
Show'rs may descend, and dews their gems disclose,
And nectar sparkle on the blooming rose.

    Such is thy pow'r, nor are thine orders vain,
O thou the leader of the mental train:
In full perfection all thy works are wrought,
And thine the sceptre o'er the realms of thought.
Before thy throne the subject-passions bow,
Of subject-passions sov'reign ruler thou;
At thy command joy rushes on the heart,
And through the glowing veins the spirits dart.

    Fancy might now her silken pinions try
To rise from earth, and sweep th' expanse on high:
From *Tithon's* bed now might *Aurora* rise,
Her cheeks all glowing with celestial dies,
While a pure stream of light o'erflows the skies.
The monarch of the day I might behold,
And all the mountains tipt with radiant gold,
But I reluctant leave the pleasing views,
Which *Fancy* dresses to delight the *Muse*;
Winter austere forbids me to aspire,
And northern tempests damp the rising fire;
They chill the tides of *Fancy's* flowing sea,
Cease then, my song, cease the unequal lay.

PHYLLIS WHEATLEY

TO THE EVENING STAR

✳

Thou fair-hair'd Angel of the Evening,
Now, whilst the sun rests on the mountains, light
Thy bright torch of love: thy radiant crown
Put on, and smile upon our evening bed!
Smile on our loves: and whilst thou drawest the
Blue curtains of the sky, scatter thy silver dew
On every flower that shuts its sweet eyes
In timely sleep. Let thy west wind sleep on
The lake: speak silence with thy glimmering eyes,
And wash the dusk with silver. Soon, full soon,
Dost thou withdraw; then the wolf rages wide,
And then the lion glares through the dun forest.
The fleeces of our flocks are covered with
Thy sacred dew: protect them with thine influence!

WILLIAM BLAKE

SONNET TO THE MOON

✳

The glitt'ring colours of the day are fled;
Come, melancholy orb! that dwell'st with night,
  Come! and o'er earth thy wand'ring lustre shed,
Thy deepest shadow, and thy softest light;
  To me congenial is the gloomy grove,
When with faint light the sloping uplands shine;
  That gloom, those pensive rays alike I love,
Whose sadness seems in sympathy with mine!
  But most for this, pale orb! thy beams are dear,
For this, benignant orb! I hail thee most:
  That while I pour the unavailing tear,
And mourn that hope to me in youth is lost,
  Thy light can visionary thoughts impart,
  And lead the Muse to soothe a suff'ring heart.

HELEN MARIA WILLIAMS

STAR-GAZERS

✳

WHAT crowd is this? what have we here! we must not pass it by;
A Telescope upon its frame, and pointed to the sky:
Long is it as a barber's pole, or mast of little boat,
Some little pleasure-skiff, that doth on Thames's waters float.

The Showman chooses well his place, 'tis Leicester's busy Square;
And is as happy in his night, for the heavens are blue and fair;
Calm, though impatient, is the crowd; each stands ready with the fee,
And envies him that's looking;—what an insight must it be!

Yet, Showman, where can lie the cause? Shall thy Implement have
    blame,
A boaster, that when he is tried, fails, and is put to shame?
Or is it good as others are, and be their eyes in fault?
Their eyes, or minds? or, finally, is yon resplendent vault?

Is nothing of that radiant pomp so good as we have here?
Or gives a thing but small delight that never can be dear?
The silver moon with all her vales, and hills of mightiest fame,
Doth she betray us when they're seen? or are they but a name?

Or is it rather that Conceit rapacious is and strong,
And bounty never yields so much but it seems to do her wrong?
Or is it, that when human Souls a journey long have had
And are returned into themselves, they cannot but be sad?

Or must we be constrained to think that these Spectators rude,
Poor in estate, of manners base, men of the multitude,
Have souls which never yet have risen, and therefore prostrate lie?
No, no, this cannot be;—men thirst for power and majesty!

Does, then, a deep and earnest thought the blissful mind employ
Of him who gazes, or has gazed? a grave and steady joy,
That doth reject all show of pride, admits no outward sign,
Because not of this noisy world, but silent and divine!

Whatever be the cause, 'tis sure that they who pry and pore
Seem to meet with little gain, seem less happy than before:
One after One they take their turn, nor have I one espied
That doth not slackly go away, as if dissatisfied.

WILLIAM WORDSWORTH

33

SONNET: TO THE AUTUMNAL MOON

✳

Mild Splendor of the various-vested Night!
Mother of wildly-working visions! hail!
I watch thy gliding, while with watery light
Thy weak eye glimmers through a fleecy veil;
And when thou lovest thy pale orb to shroud
Behind the gathered blackness lost on high;
And when thou dartest from the wind-rent cloud
Thy placid lightning o'er the awakened sky.
Ah, such is Hope! As changeful and as fair!
Now dimly peering on the wistful sight;
Now hid behind the dragon-winged Despair:
But soon emerging in her radiant might
She o'er the sorrow-clouded breast of Care
Sails, like a meteor kindling in its flight.

SAMUEL COLERIDGE

34

TO THE MOON

✳

Art thou pale for weariness
Of climbing heaven and gazing on the earth,
    Wandering companionless
Among the stars that have a different birth,—
And ever changing, like a joyless eye
That finds no object worth its constancy?

PERCY BYSSHE SHELLEY

35

THE CONSTELLATIONS

✳

O Constellations of the early night,
That sparkled brighter as the twilight died,
And made the darkness glorious! I have seen
Your rays grow dim upon the horizon's edge,
And sink behind the mountains. I have seen
The great Orion, with his jewelled belt,
That large-limbed warrior of the skies, go down
Into the gloom. Beside him sank a crowd
Of shining ones. I look in vain to find
The group of sister-stars, which mothers love
To show their wondering babes, the gentle Seven.
Along the desert space mine eyes in vain
Seek the resplendent cressets which the Twins
Uplifted in their ever-youthful hands.
The streaming tresses of the Egyptian Queen
Spangle the heavens no more. The Virgin trails
No more her glittering garments through the blue.
Gone! all are gone! and the forsaken Night,
With all her winds, in all her dreary wastes,
Sighs that they shine upon her face no more.

Now only here and there a little star
Looks forth alone. Ah me! I know them not,
Those dim successors of the numberless host
That filled the heavenly fields, and flung to earth
Their quivering fires. And now the middle watch
Betwixt the eve and morn is past, and still
The darkness gains upon the sky, and still
It closes round my way. Shall, then, the night,
Grow starless in her later hours? Have these

No train of flaming watchers, that shall mark
Their coming and farewell? O Sons of Light!
Have ye then left me ere the dawn of day
To grope along my journey sad and faint?

Thus I complained, and from the darkness round
A voice replied—was it indeed a voice,
Or seeming accents of a waking dream
Heard by the inner ear? But thus it said:
O Traveller of the Night! thine eyes are dim
With watching; and the mists, that chill the vale
Down which thy feet are passing, hide from view
The ever-burning stars. It is thy sight
That is so dark, and not the heavens. Thine eyes,
Were they but clear, would see a fiery host
Above thee; Hercules, with flashing mace,
The Lyre with silver cords, the Swan uppoised
On gleaming wings, the Dolphin gliding on
With glistening scales, and that poetic steed,
With beamy mane, whose hoof struck out from earth
The fount of Hippocrene, and many more,
Fair clustered splendors, with whose rays the Night
Shall close her march in glory, ere she yield,
To the young Day, the great earth steeped in dew.
So spake the monitor, and I perceived
How vain were my repinings, and my thought
Went backward to the vanished years and all
The good and great who came and passed with them,
And knew that ever would the years to come
Bring with them, in their course, the good and great,
Lights of the world, though, to my clouded sight,
Their rays might seem but dim, or reach me not.

WILLIAM CULLEN BRYANT

36

ON FIRST LOOKING INTO CHAPMAN'S HOMER

✳

Much have I travell'd in the realms of gold,
  And many goodly states and kingdoms seen;
  Round many western islands have I been
Which bards in fealty to Apollo hold.
Oft of one wide expanse had I been told
  That deep-brow'd Homer ruled as his demesne;
  Yet did I never breathe its pure serene
Till I heard Chapman speak out loud and bold:
Then felt I like some watcher of the skies
  When a new planet swims into his ken;
Or like stout Cortez when with eagle eyes
  He star'd at the Pacific—and all his men
Look'd at each other with a wild surmise—
  Silent, upon a peak in Darien.

JOHN KEATS

37

WINTER MORNING

✳

Sunshine and frost: a day of wonder!
My lovely friend, you're still in slumber—
It's time, my beauty, to go forth:
Open your eyes that bliss sealed tight,
To north Aurora turn your sight,
And rise, star of the north!

Last night, recall the blizzard's roar,
The dark that churned outside the door;
The moon, like a pale blotch,
Yellowed behind the somber clouds,
And there you sat in gloom and doubts—
But now... come here and watch:

Under blue skies, the sparkling snow,
Like a resplendent carpet, glows
Bright in the sun before our eyes.
The leafless woods alone are dark,
The frosted fir tree a green spark.
The river gleams under the ice.

Our whole room shimmers in soft ambers.
In the warm stove, rollicking embers
Crackle as they dance away.
It's fun to laze without a care.
But listen: let's have the brown mare
Harnessed to our sleigh.

Gliding along the morning snow,
Dear friend, how rapidly we'll go
And with the eager horse will flee
To see the empty fields' expanse,
The woods, just recently so dense,
And then the shore, so dear to me.

ALEXANDER PUSHKIN

Translation by Anton Yakovlev

38

FROM "SONG OF NATURE"

✳

Mine are the night and morning,
The pits of air, the gulf of space,
The sportive sun, the gibbous moon,
The innumerable days.

I hide in the solar glory,
I am dumb in the pealing song,
I rest on the pitch of the torrent,
In slumber I am strong.

No numbers have counted my tallies,
No tribes my house can fill,
I sit by the shining Fount of Life
And pour the deluge still;

And ever by delicate powers
Gathering along the centuries
From race on race the rarest flowers,
My wreath shall nothing miss.

And many a thousand summers
My gardens ripened well,
And light from meliorating stars
With firmer glory fell.

I wrote the past in characters
Of rock and fire the scroll,
The building in the coral sea,
The planting of the coal.

And thefts from satellites and rings
And broken stars I drew,
And out of spent and aged things
I formed the world anew;

What time the gods kept carnival,
Tricked out in star and flower,
And in cramp elf and saurian forms
They swathed their too much power.

Time and Thought were my surveyors,
They laid their courses well,
They boiled the sea, and piled the layers
Of granite, marl and shell.

But he, the man-child glorious,—
Where tarries he the while?
The rainbow shines his harbinger,
The sunset gleams his smile.

My boreal lights leap upward,
Forthright my planets roll,
And still the man-child is not born,
The summit of the whole.

Must time and tide forever run?
Will never my winds go sleep in the west?
Will never my wheels which whirl the sun
And satellites have rest?

RALPH WALDO EMERSON

39

✳

Adam. Methinks this is the zodiac of the earth,
Which rounds us with a visionary dread,—
Responding with twelve shadowy signs of earth,
In fantasque apposition and approach,
To those celestial, constellated twelve
Which palpitate adown the silent nights
Under the pressure of the hand of God
Stretched wide in benediction. At this hour,
Not a star pricketh the flat gloom of heaven!
But, girdling close our nether wilderness,
The zodiac-figures of the earth loom slow,—
Drawn out, as suiteth with the place and time,
In twelve colossal shades instead of stars,
Through which the ecliptic line of mystery
Strikes bleakly with an unrelenting scope,
Foreshowing life and death.

ELIZABETH BARRETT BROWNING

40

THE LIGHT OF STARS

✴

The night is come, but not too soon;
  And sinking silently,
All silently, the little moon
  Drops down behind the sky.

There is no light in earth or heaven
  But the cold light of stars;
And the first watch of night is given
  To the red planet Mars.

Is it the tender star of love?
  The star of love and dreams?
O no! from that blue tent above,
  A hero's armor gleams.

And earnest thoughts within me rise,
  When I behold afar,
Suspended in the evening skies,
  The shield of that red star.

O star of strength! I see thee stand
  And smile upon my pain;
Thou beckonest with thy mailèd hand,
  And I am strong again.

Within my breast there is no light
  But the cold light of stars;
I give the first watch of the night
  To the red planet Mars.

The star of the unconquered will,
  He rises in my breast,
Serene, and resolute, and still,
  And calm, and self-possessed.

And thou, too, whosoe'er thou art,
  That readest this brief psalm,
As one by one thy hopes depart,
  Be resolute and calm.

O fear not in a world like this,
  And thou shalt know erelong,
Know how sublime a thing it is
  To suffer and be strong.

HENRY WADSWORTH LONGFELLOW

41

FROM "AL AARAAF"

✳

What tho' in worlds which sightless cycles run,
Link'd to a little system, and one sun—
Where all my love is folly and the crowd
Still think my terrors but the thunder cloud,
The storm, the earthquake, and the ocean-wrath
(Ah! will they cross me in my angrier path?)—
What tho' in worlds which own a single sun
The sands of Time grow dimmer as they run,
Yet thine is my resplendency, so given
To bear my secrets thro' the upper Heaven.
Leave tenantless thy crystal home, and fly,
With all thy train, athwart the moony sky—
Apart—like fire-flies in Sicilian night,
And wing to other worlds another light!
Divulge the secrets of thy embassy
To the proud orbs that twinkle—and so be
To ev'ry heart a barrier and a ban
Lest the stars totter in the guilt of man!

EDGAR ALLAN POE

Poe set "Al Aaraaf" in the supernova discovered by Tycho Brahe in 1572.

42

FROM "TIMBUCTOO"

✳

I felt my soul grow mighty, and my spirit
With supernatural excitation bound
Within me, and my mental eye grew large
With such a vast circumference of thought,
That in my vanity I seem'd to stand
Upon the outward verge and bound alone
Of full beatitude. Each failing sense
As with a momentary flash of light
Grew thrillingly distinct and keen. I saw
The smallest grain that dappled the dark Earth,
The indistinctest atom in deep air,
The Moon's white cities, and the opal width
Of her small glowing lakes, her silver heights
Unvisited with dew of vagrant cloud,
And the unsounded, undescended depth
Of her black hollows. The clear Galaxy
Shorn of its hoary lustre, wonderful,
Distinct and vivid with sharp points of light
Blaze within blaze, an unimagin'd depth
And harmony of planet-girded Suns
And moon-encircled planets, wheel in wheel,
Arch'd the wan Sapphire. Nay, the hum of men,
Or other things talking in unknown tongues,
And notes of busy life in distant worlds
Beat like a far wave on my anxious ear.

ALFRED, LORD TENNYSON

43

LOVE

✳

We two that planets erst had been
Are now a double star,
And in the heavens may be seen,
Where that we fixèd are.

Yet, whirled with subtle power along,
Into new space we enter,
And evermore with spheral song
Revolve about one centre.

HENRY DAVID THOREAU

44

WHEN I HEARD THE LEARN'D ASTRONOMER

✳

When I heard the learn'd astronomer,
When the proofs, the figures, were ranged in columns before me,
When I was shown the charts and diagrams, to add, divide, and
    measure them,
When I sitting heard the astronomer where he lectured with much
    applause in the lecture-room,
How soon unaccountable I became tired and sick,
Till rising and gliding out I wander'd off by myself,
In the mystical moist night-air, and from time to time,
Look'd up in perfect silence at the stars.

WALT WHITMAN

45

SORROWS OF THE MOON

✳

The Moon more indolently dreams to-night
Than a fair woman on her couch at rest.
Caressing, with a hand distraught and light,
Before she sleeps, the contour of her breast.

Upon her silken avalanche of down,
Dying she breathes a long and swooning sigh;
And watches the white visions past her flown,
Which rise like blossoms to the azure sky.

And when, at times, wrapped in her languor deep,
Earthward she lets a furtive tear-drop flow,
Some pious poet, enemy of sleep,

Takes in his hollow hand the tear of snow
Whence gleams of iris and of opal start,
And hides it from the Sun, deep in his heart.

CHARLES BAUDELAIRE

Translation by Frank Pearce Sturm

46

THE PHILOSOPHER AND THE STARS

✳

And you, ye stars,
Who slowly begin to marshal,
As of old, in the fields of heaven,
Your distant, melancholy lines—
Have you, too, survived yourselves?
Are you, too, what I fear to become?
You, too, once lived—
You too moved joyfully
Among august companions
In an older world, peopled by Gods,
In a mightier order,
The radiant, rejoicing, intelligent Sons of Heaven!
But now, you kindle
Your lonely, cold-shining lights,
Unwilling lingerers
In the heavenly wilderness,
For a younger, ignoble world.
And renew, by necessity,
Night after night your courses,
In echoing unnear'd silence,
Above a race you know not.
Uncaring and undelighted,
Without friend and without home.
Weary like us, though not
Weary with our weariness.

MATTHEW ARNOLD

47

A GRAIN OF SAND

✳

Do you see this grain of sand
Lying loosely in my hand?
Do you know to me it brought
Just a simple loving thought?
When one gazes night by night
On the glorious stars of light,
Oh how little seems the span
Measured round the life of man.

Oh! how fleeting are his years
With their smiles and their tears;
Can it be that God does care
For such atoms as we are?
Then outspake this grain of sand
"I was fashioned by His hand
In the star lit realms of space
I was made to have a place.

"Should the ocean flood the world,
Were its mountains 'gainst me hurled
All the force they could employ
Wouldn't a single grain destroy;
And if I, a thing so light,
Have a place within His sight;
You are linked unto his throne
Cannot live nor die alone.

In the everlasting arms
Mid life's dangers and alarms
Let calm trust your spirit fill;
Know He's God, and then be still."
Trustingly I raised my head
Hearing what the atom said;
Knowing man is greater far
Than the brightest sun or star.

FRANCES ELLEN WATKINS HARPER

48

TO A STAR SEEN AT TWILIGHT

✳

Hail solitary star!
That shinest from thy far blue height,
And overlookest Earth
And Heaven, companionless in light!
The rays around thy brow
Are an eternal wreath for thee;
Yet thou'rt not proud, like man,
Though thy broad mirror is the sea,
And thy calm home eternity!

Shine on, night-bosomed star!
And through its realms thy soul's eye dart,
And count each age of light,
For their eternal wheel thou art.

Thou dost roll into the past days,
Years, and ages too,
And naught thy giant progress stays.

I love to gaze upon
Thy speaking face, thy calm, fair brow,
And feel my spirit dark
And deep, grow bright and pure as thou.
Like thee it stands alone:
Like thee its native home is night,
But there the likeness ends,—
It beams not with thy steady light.
Its upward path is high,
But not so high as thine—thou'rt far

Above the reach of clouds,
Of storms, of wreck, oh lofty star!
I would all men might look
Upon thy pure sublimity,
And in their bosoms drink
Thy lovliness and light like me;
For who in all the world
Could gaze upon thee thus, and feel
Aught in his nature base,
Or mean, or low, around him steal!

Shine on companionless
As now thou seem'st. Thou art the throne
Of thy own spirit, star!
And mighty things must be alone.
Alone the ocean heaves,
Or calms his bosom into sleep;
Alone each mountain stands
Upon its basis broad and deep;
Alone through heaven the comets sweep,
Those burning worlds which God has thrown
Upon the universe in wrath,
As if he hated them—their path
No stars, no suns may follow, *none*—
'T is great, 't is great to be alone!

JOHN ROLLIN RIDGE

49

EMIGRAVIT

✳

With sails full set, the ship her anchor weighs.
Strange names shine out beneath her figure head.
What glad farewells with eager eyes are said!
What cheer for him who goes, and him who stays!
Fair skies, rich lands, new homes, and untried days
Some go to seek: the rest but wait instead,
Watching the way wherein their comrades led,
Until the next stanch ship her flag doth raise.
Who knows what myriad colonies there are
Of fairest fields, and rich, undreamed-of gains
Thick planted in the distant shining plains
Which we call sky because they lie so far?
Oh, write of me, not "Died in bitter pains,"
But "Emigrated to another star!"

HELEN HUNT JACKSON

50

THE MOTHER MOON

✳

The moon upon the wide sea
Placidly looks down,
Smiling with her mild face,
Though the ocean frown.
Clouds may dim her brightness,
But soon they pass away,
And she shines out, unaltered,
O'er the little waves at play.
So 'mid the storm or sunshine,
Wherever she may go,
Led on by her hidden power
The wild sea must plow.

As the tranquil evening moon
Looks on that restless sea,
So a mother's gentle face,
Little child, is watching thee.
Then banish every tempest,
Chase all your clouds away,
That smoothly and brightly
Your quiet heart may play.
Let cheerful looks and actions
Like shining ripples flow,
Following the mother's voice,
Singing as they go.

LOUISA MAY ALCOTT

I KNOW NOT WHAT I SEEK ETERNALLY

✳

I know not what I seek eternally
on earth, in the air, and in the sky;
I know not what I seek: something I lost
I know not when and can no longer find,
even when dreaming that, invisible,
it lives in all I touch and all I see.
Happiness, I shall not find you ever
on earth, in the air, or in the sky;
   although I know you're real
   and not an empty dream!

ROSALÍA DE CASTRO

Translation by Rhina P. Espaillat

52

THE OLD ASTRONOMER

✳

REACH me down my Tycho Brahé,—I would know him when we
   meet,
When I share my later science, sitting humbly at his feet;
He may know the law of all things, yet be ignorant of how
We are working to completion, working on from then till now.

Pray, remember, that I leave you all my theory complete,
Lacking only certain data, for your adding, as is meet;
And remember, men will scorn it, 'tis original and true,
And the obloquy of newness may fall bitterly on you.

But, my pupil, as my pupil you have learnt the worth of scorn;
You have laughed with me at pity, we have joyed to be forlorn;
What, for us, are all distractions of men's fellowship and smiles?
What, for us, the goddess Pleasure, with her meretricious wiles?

You may tell that German college that their honour comes too late.
But they must not waste repentance on the grizzly savant's fate;
Though my soul may set in darkness, it will rise in perfect light;
I have loved the stars too truly to be fearful of the night.

What, my boy, you are not weeping? You should save your eyes for
   sight;
You will need them, mine observer, yet for many another night.
I leave none but you, my pupil, unto whom my plans are known.
You "have none but me," you murmur, and I "leave you quite alone"?

Well then, kiss me,—since my mother left her blessing on my brow,
There has been a something wanting in my nature until now;
I can dimly comprehend it,—that I might have been more kind,
Might have cherished you more wisely, as the one I leave behind.

I "have never failed in kindness"? No, we lived too high for strife,—
Calmest coldness was the error which has crept into our life;
But your spirit is untainted, I can dedicate you still
To the service of our science: you will further it? you will!

There are certain calculations I should like to make with you,
To be sure that your deductions will be logical and true;
And remember, "Patience, Patience," is the watchword of a sage,
Not to-day nor yet to-morrow can complete a perfect age.

I have sown, like Tycho Brahé, that a greater man may reap;
But if none should do my reaping, 'twill disturb me in my sleep.
So be careful and be faithful, though, like me, you leave no name;
See, my boy, that nothing turn you to the mere pursuit of fame.

I must say Good-bye, my pupil, for I cannot longer speak;
Draw the curtain back for Venus, ere my vision grows too weak:
It is strange the pearly planet should look red as fiery Mars,—
God will mercifully guide me on my way amongst the stars.

SARAH WILLIAMS

53

WAITING BOTH

✳

A star looks down at me,
And says: "Here I and you
Stand, each in our degree.
What do you mean to do,—
      Mean to do?"

 I say: "For all I know,
Wait, and let Time go by,
Till my change come."—"Just so."
The star says: "So mean I:—
      So mean I."

THOMAS HARDY

54

" — I AM LIKE A SLIP OF COMET "

✳

        — I am like a slip of comet,
Scarce worth discovery, in some corner seen
Bridging the slender difference of two stars,
Come out of space, or suddenly engender'd
By heady elements, for no man knows:
But when she sights the sun she grows and sizes
And spins her skirts out, while her central star
Shakes its cocooning mists; and so she comes
To fields of light; millions of travelling rays
Pierce her; she hangs upon the flame-cased sun,
And sucks the light as full as Gideon's fleece:
But then her tether calls her; she falls off,
And as she dwindles shreds her smock of gold
Amidst the sistering planets, till she comes
To single Saturn, last and solitary;
And then goes out into the cavernous dark.
So I go out: my little sweet is done:
I have drawn heat from this contagious sun:
To not ungentle death now forth I run.

GERARD MANLEY HOPKINS

A SOLAR ECLIPSE

✳

In that great journey of the stars through space
    About the mighty, all-directing Sun,
    The pallid, faithful Moon has been the one
Companion of the Earth. Her tender face,
Pale with the swift, keen purpose of that race
    Which at Time's natal hour was first begun,
    Shines ever on her lover as they run
And lights his orbit with her silvery smile.

Sometimes such passionate love doth in her rise,
    Down from her beaten path she softly slips,
And with her mantle veils the Sun's bold eyes,
    Then in the gloaming finds her lover's lips.
While far and near the men our world call wise
    See only that the Sun is in eclipse.

ELLA WHEELER WILCOX

56

STARS, I HAVE SEEN THEM FALL

✳

Stars, I have seen them fall,
  But when they drop and die
No star is lost at all
  From all the star-sown sky.
The toil of all that be
  Helps not the primal fault;
It rains into the sea,
  And still the sea is salt.

A. E. HOUSMAN

57

EARTH

✴

If this little world to-night
  Suddenly should fall through space
In a hissing headlong flight,
  Shrivelling from off its face,
As it falls into the sun,
  In an instant every trace
  Of the little crawling things—
  Ants, philosophers, and lice,

Cattle, cockroaches, and kings,
  Beggars, millionaires, and mice,
Men and maggots all as one
  As it falls into the sun. . . .
Who can say but at the same
  Instant from some planet far
A child may watch us and exclaim:
  "See the pretty shooting star!"

OLIVER HERFORD

58

STARLIGHT

✳

O beautiful stars, when you see me go
  Hither and thither in search of love,
Do you think me faithless, who gleam and glow
  Serene and fixed in the blue above?
    O stars, so golden, it is not so.

But there is a garden I dare not see,
  There is a place where I fear to go,
Since the charm and glory of life to me
  The brown earth covered there long ago.
    O stars, you saw it, you know, you know.

Hither and thither I wandering go,
  With aimless haste and wearying fret;
In a search for pleasure and love? Not so,
  Seeking desperately to forget.
    You see so many, O stars, you know.

LAURENCE HOPE (ADELA FLORENCE CORY NICOLSON)

THE CAT AND THE MOON

✳

The cat went here and there
And the moon spun round like a top,
And the nearest kin of the moon,
The creeping cat, looked up.
Black Minnaloushe stared at the moon,
For, wander and wail as he would,
The pure cold light in the sky
Troubled his animal blood.
Minnaloushe runs in the grass
Lifting his delicate feet.
Do you dance, Minnaloushe, do you dance?
When two close kindred meet,
What better than call a dance?
Maybe the moon may learn,
Tired of that courtly fashion,
A new dance turn.
Minnaloushe creeps through the grass
From moonlit place to place,
The sacred moon overhead
Has taken a new phase.
Does Minnaloushe know that his pupils
Will pass from change to change,
And that from round to crescent,
From crescent to round they range?

Minnaloushe creeps through the grass
Alone, important and wise,
And lifts to the changing moon
His changing eyes.

WILLIAM BUTLER YEATS

60

ALFONSO CHURCHILL

✳

They laughed at me as "Prof. Moon,"
As a boy in Spoon River, born with the thirst
Of knowing about the stars.
They jeered when I spoke of the lunar mountains,
And the thrilling heat and cold,
And the ebon valleys by silver peaks,
And Spica quadrillions of miles away,
And the littleness of man.
But now that my grave is honored, friends,
Let it not be because I taught
The lore of the stars in Knox College,
But rather for this: that through the stars
I preached the greatness of man,
Who is none the less a part of the scheme of things
For the distance of Spica or the Spiral Nebulae;
Nor any the less a part of the question
Of what the drama means.

EDGAR LEE MASTERS

61

TO HALLEY'S COMET

✳

Thou "Wanderer" out in the vast Unknown,
      When next upon thy path around the sun
      Thou dost return, how many will have run
Their race who saw thee last, and youth have grown
To age, and many changes will be here!
      Such progress has the mind of man achieved
      In knowledge of the heavens, 'tis scarce believed;
Yet more 'twill know when next thou shalt appear.
While thou, returning, wilt thy path pursue
      For centuries, and many millions will
      Thy coming watch and recognize, until,
At last thou, too, shalt disappear from view;—
      Worn out, dissolved, and scattered far through space,
      No more shall men behold thee in thy place!

ALICE BERLINGETT

62

A SONG OF A NAVAJO WEAVER

✳

For ages long, my people have been
   Dwellers in this land;
For ages viewed these mountains,
   Loved these mesas and these sands,
That stretch afar and glisten,
   Glimmering in the sun
As it lights the mighty canons
   Ere the weary day is done.
Shall I, a patient dweller in this
   Land of fair blue skies,
Tell something of their story while
   My shuttle swiftly flies?
As I weave I'll trace their journey,
   Devious, rough and wandering,
Ere they reached the silent region
   Where the night stars seem to sing.
When the myriads of them glitter
   Over peak and desert waste,
Crossing which the silent runner and
   The gaunt of co-yo-tees haste.
Shall I weave the zig-zag pathway
   Whence the sacred fire was born;
And interweave the symbol of the God
   Who brought the corn—
Of the Rain-god whose fierce anger
   Was appeased by sacred meal,
And the trust that my brave people
   In him evermore shall feel?

All this perhaps I might weave
   As the woof goes to and fro,
Wafting as my shuttle passes,
   Humble hopes, and joys and care,
Weaving closely, weaving slowly,
   While I watch the pattern grow;
Showing something of my life:
   To the Spirit God a prayer.
Grateful that he brought my people
   To the land of silence vast
Taught them arts of peace and ended
   All their wanderings of the past.
Deftly now I trace the figures,
   This of joy and that of woe;
And I leave an open gate-way
   For the Dau to come and go.

BERTRAND N. O. WALKER (HEN-TAH)

63

✳

A cloud fell down from the heavens,
    And broke on the mountain's brow;
It scattered the dusky fragments
    All over the vale below.

The moon and the stars were anxious
    To know what its fate might be;
So they rushed to the azure op'ning,
    And all peered down to see.

PAUL LAURENCE DUNBAR

64

THE STAR-SPLITTER

✳

"You know Orion always comes up sideways.
Throwing a leg up over our fence of mountains,
And rising on his hands, he looks in on me
Busy outdoors by lantern-light with something
I should have done by daylight, and indeed,
After the ground is frozen, I should have done
Before it froze, and a gust flings a handful
Of waste leaves at my smoky lantern chimney
To make fun of my way of doing things,
Or else fun of Orion's having caught me.
Has a man, I should like to ask, no rights
These forces are obliged to pay respect to?"
So Brad McLaughlin mingled reckless talk
Of heavenly stars with hugger-mugger farming,
Till having failed at hugger-mugger farming,
He burned his house down for the fire insurance
And spent the proceeds on a telescope
To satisfy a lifelong curiosity
About our place among the infinities.

"What do you want with one of those blame things?"
I asked him well beforehand. "Don't you get one!"

"Don't call it blamed; there isn't anything
More blameless in the sense of being less
A weapon in our human fight," he said.
"I'll have one if I sell my farm to buy it."
There where he moved the rocks to plow the ground
And plowed between the rocks he couldn't move,

Few farms changed hands; so rather than spend years
Trying to sell his farm and then not selling,
He burned his house down for the fire insurance
And bought the telescope with what it came to.
He had been heard to say by several:
"The best thing that we're put here for's to see;
The strongest thing that's given us to see with's
A telescope. Someone in every town
Seems to me owes it to the town to keep one.
In Littleton it may as well be me."
After such loose talk it was no surprise
When he did what he did and burned his house down.

Mean laughter went about the town that day
To let him know we weren't the least imposed on,
And he could wait—we'd see to him tomorrow.
But the first thing next morning we reflected
If one by one we counted people out
For the least sin, it wouldn't take us long
To get so we had no one left to live with.
For to be social is to be forgiving.
Our thief, the one who does our stealing from us,
We don't cut off from coming to church suppers,
But what we miss we go to him and ask for.
He promptly gives it back, that is if still
Uneaten, unworn out, or undisposed of.
It wouldn't do to be too hard on Brad
About his telescope. Beyond the age
Of being given one for Christmas gift,
He had to take the best way he knew how
To find himself in one. Well, all we said was
He took a strange thing to be roguish over.
Some sympathy was wasted on the house,

A good old-timer dating back along;
But a house isn't sentient; the house
Didn't feel anything. And if it did,
Why not regard it as a sacrifice,
And an old-fashioned sacrifice by fire,
Instead of a new-fashioned one at auction?

Out of a house and so out of a farm
At one stroke (of a match), Brad had to turn
To earn a living on the Concord railroad,
As under-ticket-agent at a station
Where his job, when he wasn't selling tickets,
Was setting out up track and down, not plants
As on a farm, but planets, evening stars
That varied in their hue from red to green.

He got a good glass for six hundred dollars.
His new job gave him leisure for stargazing.
Often he bid me come and have a look
Up the brass barrel, velvet black inside,
At a star quaking in the other end.
I recollect a night of broken clouds
And underfoot snow melted down to ice,
And melting further in the wind to mud.
Bradford and I had out the telescope.
We spread our two legs as it spread its three,
Pointed our thoughts the way we pointed it,
And standing at our leisure till the day broke,
Said some of the best things we ever said.
That telescope was christened the Star-Splitter,
Because it didn't do a thing but split
A star in two or three the way you split
A globule of quicksilver in your hand
With one stroke of your finger in the middle.

It's a star-splitter if there ever was one,
And ought to do some good if splitting stars
'Sa thing to be compared with splitting wood.

We've looked and looked, but after all where are we?
Do we know any better where we are,
And how it stands between the night tonight
And a man with a smoky lantern chimney?
How different from the way it ever stood?

ROBERT FROST

65

STARS

✳

Alone in the night
  On a dark hill
With pines around me
  Spicy and still,

And a heaven full of stars
  Over my head,
White and topaz
  And misty red;

Myriads with beating
  Hearts of fire
That aeons
  Cannot vex or tire;

Up the dome of heaven
  Like a great hill,
I watch them marching
  Stately and still,

And I know that I
  Am honored to be
Witness
  Of so much majesty.

SARA TEASDALE

66

THE SONG OF THE STARS

✳

We are the stars which sing,
We sing with our light;
We are the birds of fire,
We fly over the sky.
Our light is a voice;
We make a road for spirits,
For the spirits to pass over.
Among us are three hunters
Who chase a bear;
There never was a time
When they were not hunting.
We look down on the mountains.
This is the Song of the Stars.

UNKNOWN

Translation from the Passamaquoddy, an Algonquian language of eastern Maine and New Brunswick, by Charles Godfrey Leland

The date of this poem is unknown. It was published in translation in the late nineteenth century.

67

A SONG OF THE MOON

✳

The moonlight breaks upon the city's domes,
And falls along cemented steel and stone,
Upon the grayness of a million homes,
Lugubrious in unchanging monotone.
Upon the clothes behind the tenement,
That hang like ghosts suspended from the lines,
Linking each flat to each indifferent,
Incongruous and strange the moonlight shines.

There is no magic from your presence here,
Ho, moon, sad moon, tuck up your trailing robe,
Whose silver seems antique and so severe
Against the glow of one electric globe.

Go spill your beauty on the laughing faces
Of happy flowers that bloom a thousand hues,
Waiting on tiptoe in the wilding spaces,
To drink your wine mixed with sweet drafts of dews.

CLAUDE MCKAY

68

✳

All I could see from where I stood
Was three long mountains and a wood;
I turned and looked another way,
And saw three islands in a bay.
So with my eyes I traced the line
Of the horizon, thin and fine,
Straight around till I was come
Back to where I'd started from;
And all I saw from where I stood
Was three long mountains and a wood.
Over these things I could not see;
These were the things that bounded me;
And I could touch them with my hand,
Almost, I thought, from where I stand.
And all at once things seemed so small
My breath came short, and scarce at all.
But, sure, the sky is big, I said;
Miles and miles above my head;
So here upon my back I'll lie
And look my fill into the sky.
And so I looked, and, after all,
The sky was not so very tall.
The sky, I said, must somewhere stop,
And—sure enough!—I see the top!
The sky, I thought, is not so grand;
I 'most could touch it with my hand!
And reaching up my hand to try,
I screamed to feel it touch the sky.

I screamed, and—lo!—Infinity
Came down and settled over me;
Forced back my scream into my chest,
Bent back my arm upon my breast,
And, pressing of the Undefined
The definition on my mind,
Held up before my eyes a glass
Through which my shrinking sight did pass
Until it seemed I must behold
Immensity made manifold;
Whispered to me a word whose sound
Deafened the air for worlds around,
And brought unmuffled to my ears
The gossiping of friendly spheres,
The creaking of the tented sky,
The ticking of Eternity.
I saw and heard, and knew at last
The How and Why of all things, past,
And present, and forevermore.
The Universe, cleft to the core,
Lay open to my probing sense
That, sick'ning, I would fain pluck thence
But could not,—nay! But needs must suck
At the great wound, and could not pluck
My lips away till I had drawn
All venom out.—Ah, fearful pawn!
For my omniscience paid I toll
In infinite remorse of soul.

EDNA ST. VINCENT MILLAY

69

BALLAD OF THE MOON, MOON

✳

The moon has come to the smithy
with its lily stowaways.
The little boy looks and looks at her;
the child is looking, amazed.
In the air ceaselessly turning,
the moon bares her arms to raise
and reveal her hard tin breasts,
so lewd and pure, to his gaze.
"Run away, O Moon, Moon, Moon.
Were Gypsies to come home, they
would craft white rings and necklaces
from the stuff of your heart this day."
"Child, you must let me dance.
When the Gypsies come, I say
they will find you on the anvil
with your small eyes locked away."

"Run away, O Moon, Moon, Moon,
for I hear their hoofbeats now."
"Leave me, child, and do not trample
the whiteness of my starched gown."
The rider was fast approaching,
beating the drum of the ground.
There in the smithy lay the child,
with his little eyelids down.

Through the olive-grove they came,
the Gypsies, all bronze and dreaming.
They rode with their heads held high
and their eyes half-closed and gleaming.

How the *zumaya* is singing
from the tree, O hear her sing!
Across the sky goes the Moon, Moon,
and a child is following.

In the smithy they are weeping,
the Gypsies wail and keen.
The air is watching, witnessing,
the air is watching the scene.

FEDERICO GARCÍA LORCA

Translation by Rhina P. Espaillat

70

CHANT

✳

In the time when Dendid created all things,
He created the sun,
And the sun is born, and dies, and comes again;
He created the moon,
And the moon is born, and dies, and comes again;
He created the stars,
And the stars are born, and die, and come again;
He created man,
And the man is born, and dies, and never comes again.

UNKNOWN, DINKA NILOTIC SUDAN

Translation by Willard Trask

The nature of the Dinka poems involves passing them from generation to generation over many, many years, so the date of this poem is unknown. It was recorded for translation in the early twentieth century.

WITNESSING THE LAUNCH OF THE SHUTTLE ATLANTIS

✳

So much of life in the world is waiting, that
This day was no exception, so we waited
All morning long and into the afternoon.
I spent some of the time remembering
Dante, who did the voyage in the mind
Alone, with no more nor heavier machinery
Than the ghost of a girl giving him guidance;

And wondered if much was lost to gain all this
New world of engine and energy, where dream
Translates into deed. But when the thing went up
It was indeed impressive, as if hell
Itself opened to send its emissary
In search of heaven or "the unpeopled world"
(thus Dante of doomed Ulysses) "behind the sun."

So much of life in the world is memory
That the moment of the happening itself—
So much with noise and smoke and rising clear
To vanish at the limit of our vision
Into the light blue light of afternoon—
Appeared no more, against the void in aim,
Than the flare of a match in sunlight, quickly snuffed.

What yet may come of this? We cannot know.
Great things are promised, as the promised land
Promised to Moses that he would not see
But a distant sight of, though the children would.
The world is made of pictures of the world,
And the pictures change the world into another world
We cannot know, as we knew not this one.

HOWARD NEMEROV

72

THE FIRST MEN ON MERCURY

✳

— We come in peace from the third planet.
Would you take us to your leader?

— Bawr stretter! Bawr. Bawr. Stretterhawl?

— This is a little plastic model
of the solar system, with working parts.
You are here and we are there and we
are now here with you, is this clear?

— Gawl horrop. Bawr Abawrhannahanna!

— Where we come from is blue and white
with brown, you see we call the brown
here 'land', the blue is 'sea', and the white
is 'clouds' over land and sea, we live
on the surface of the brown land,
all round is sea and clouds. We are 'men'.
Men come —

— Glawp men! Gawrbenner menko. Menhawl?

— Men come in peace from the third planet
which we call 'earth'. We are earthmen.
Take us earthmen to your leader.

— Thmen? Thmen? Bawr. Bawrhossop.
Yuleeda tan hanna. Harrabost yuleeda.

— I am the yuleeda. You see my hands,
we carry no benner, we come in peace.
The spaceways are all stretterhawn.

— Glawn peacemen all horrabhanna tantko!
Tan come at'mstrossop. Glawp yuleeda!

— Atoms are peacegawl in our harraban.
Menbat worrabost from tan hannahanna.

— You men we know bawrhossoptant. Bawr.
We know yuleeda. Go strawg backspetter quick.

— We cantantabawr, tantingko backspetter now!

— Banghapper now! Yes, third planet back.
Yuleeda will go back blue, white, brown
nowhanna! There is no more talk.

— Gawl han fasthapper?

— No. You must go back to your planet.
Go back in peace, take what you have gained
but quickly.

— Stretterworra gawl, gawl...

— Of course, but nothing is ever the same,
now is it? You'll remember Mercury.

EDWIN MORGAN

73

IN THE FIELD

✳

This field-grass brushed our legs
Last night, when out we stumbled looking up,
Wading as through the cloudy dregs
Of a wide, sparkling cup,

Our thrown-back heads aswim
In the grand, kept appointments of the air,
Save where a pine at the sky's rim
Took something from the Bear.

Black in her glinting chains,
Andromeda feared nothing from the seas,
Preserved as by no hero's pains,
Or hushed Euripides',

And there the dolphin glowed,
Still flailing through a diamond froth of stars,
Flawless as when Arion rode
One of its avatars.

But none of that was true,
What shapes that Greece or Babylon discerned
Had time not slowly drawn askew
Or like cat's cradles turned?

And did we not recall
That Egypt's north was in the Dragon's tail?
As if a form of type should fall
And dash itself like hail,

The heavens jumped away,
Bursting the cincture of the zodiac,
    Shot flares with nothing left to say
        To us, not coming back

    Unless they should at last,
Like hard-flung dice that ramble out the throw,
    Be gathered for another cast.
        Whether that might be so

    We could not say, but trued
Our talk awhile to words of the real sky,
    Chatting of class or magnitude,
        Star-clusters, nebulae,

    And how Antares, huge
As Mars' big roundhouse swing, and more, was fled
    As in some rimless centrifuge
        Into a blink of red.

    It was the nip of fear
That told us when imagination caught
    The feel of what we said, came near
        The schoolbook thoughts we thought,

    And faked a scan of space
Blown black and hollow by our spent grenade,
    All worlds dashed out without a trace,
        The very light unmade.

    Then, in the late-night chill,
We turned and picked our way through outcrop stone
    By the faint starlight, up the hill
        To where our bed-lamp shone.

Today, in the same field,
The sun takes all, and what could lie beyond?
Those holes in heaven have been sealed
Like rain-drills in a pond,

And we, beheld in gold,
See nothing starry but these galaxies
Of flowers, dense and manifold,
Which lift about our knees—

White daisy-drifts where you
Sink down to pick an armload as we pass,
Sighting the heal-all's minor blue
In chasms of the grass,

And strews of hawkweed where,
Amongst the reds or yellows as they burn,
A few dead polls commit to air
The seeds of their return.

We could no doubt mistake
These flowers for some answers to that fright
We felt for all creation's sake
In our dark talk last night,

Taking to heart what came
Of the heart's wish for life, which, staking here
In the least field an endless claim,
Beats on from sphere to sphere

And pounds beyond the sun,
Where nothing less peremptory can go,
And is ourselves, and is the one
Unbounded thing we know.

RICHARD WILBUR

74

HIGH FLIGHT

✴

Oh! I have slipped the surly bonds of Earth
And danced the skies on laughter-silvered wings;
Sunward I've climbed, and joined the tumbling mirth
Of sun-split clouds, — and done a hundred things
You have not dreamed of — wheeled and soared and swung
High in the sunlit silence. Hov'ring there,
I've chased the shouting wind along, and flung
My eager craft through footless halls of air.....

Up, up the long, delirious burning blue
I've topped the wind-swept heights with easy grace
Where never lark, or ever eagle flew —
And, while with silent, lifting mind I've trod
The high untrespassed sanctity of space,
Put out my hand, and touched the face of God.

JOHN GILLESPIE MAGEE, JR.

75

DELAY

✳

The radiance of the star that leans on me
Was shining years ago. The light that now
Glitters up there my eyes may never see,
And so the time lag teases me with how

Love that loves now may not reach me until
Its first desire is spent. The star's impulse
Must wait for eyes to claim it beautiful
And love arrived may find us somewhere else.

ELIZABETH JENNINGS

76

LADDER TO THE MOON

✳

If I had a ladder that reached to the moon
Up its millions of rungs I would go,
Up higher than ever the clouds can fly
Till the earth was a ball below.

I'd put on my warm wool winter coat
And my long scarlet scarf in case
While I climbed my ladder right up to the moon
It should start to snow in space.

I'd sidestep a couple of shooting stars
I'd stand on the steepest hill
At the top of my ladder to the moon
If only the moon stood still.

X. J. KENNEDY

77

FOR MY GREAT-GREAT GRANDSON THE SPACE PIONEER

✳

You, What's-your-name, who down the byways of my blood
are hurtling toward the future, tell me if you've packed
the thousand flavors of the wind, the river's voice,
the tongues of moss and fern singing the earth.

And where have you left the rain? Careful: don't lose it,
nor the moan of the seagull in her blue desert,
nor those stars warm as caresses
you will not find again in your nights of steel.

Watch that you don't run short of butterflies;
learn the colors of the hours;
and here, in this little case of bones
I've left you the perfume of the sea.

PARA MI TATARANIETO EL ASTROPIONERO

Tú, Fulanito, que por los caminos de mi sangre
te lanzas al futuro, dime si te llevas
los mil sabores del viento, la voz del río,
las lenguas de musgo y helecho que cantan la tierra.

Y dónde dejaste la lluvia? Que no se te pierda,
ni el gemir de la gaviota en su desierto azul,
ni esas estrellas tibias como caricias
que no encontrarás en tus noches de acero.

Fíjate que no te falten mariposas;
apréndete el color de las horas;
y toma, que en esta cajita de huesos
te dejo el perfume de los mares.

RHINA P. ESPAILLAT

78

ECLIPSE

✳

Moon, half rusted away
in the sun's indomitable shadow,

I stand at the frosted window
wrapped in a flannel robe

and see not what Galileo saw—
a universe of planets spinning

like plates from the hands
of a master juggler—

but you, our one moon,
slender at times,

at times full as a breast
brimming with milky light.

If the sun is a warrior
in flaming armor,

the moon is a ghost
disappearing,

leaving behind
the merest trace of stars.

LINDA PASTAN

79

GALAXIES

✳

What is a galaxy to me
compared with morning dew in sun
that glows and sparkles? I can see
the question is the same. Just one
unsolved conundrum. Tell me why
the maker's maker is unknown
and yet each one of us must die?
Chaos and cosmos cannot bring
the answer. Dolphins, birds and bats
can navigate. My golden ring
comes from a star. The eyes of cats
have understanding but evade
my searching gaze. I will not let
this tyrant life make me afraid.
I understand and then forget.
The solar winds play through the hair
of planets—trombones twine with flutes
and darkness flirts with infrared,
beyond my spectrum colour shoots
pinheads of brilliance. Someone said
the world is in a grain of sand
and as the telescopes display
God's grains of grit they seem less grand
than human love. I dare to say

that everything vibrates with bliss.
Already I am full of awe;
no greater wonder here than this
adventure. There is nothing more.

JANET KENNY

80

THE BALLET OF THE EIGHT-WEEK KITTENS

✳

With such abandon—buoyant wide *jetés*
around the kitchen, furniture and air
possessed by arcs of fur and ricochets,
plunges and cabrioles—they're everywhere.
Dance, kittens. Take the table, flying
jump-drunk: You have cause to pirouette,
more than you dream of—barns and meadows lying
outside—things you don't imagine yet.
Hurl yourselves in knots across the floor;
leaps, demented *entrechats*. Ignore
the galaxies beyond the kitchen door.
And, when you tumble to it that there's *more*
than this—more than the little world you know—
take me out there with you when you go.

DEBORAH WARREN

81

RAINY ECLIPSE

*

for Mario and Martha Joy Gottfried
Valle de Bravo, Mexico. July 11, 1991

A starling, like a piece of chipped basalt,
Swerves high above the dark volcano, quiet
As ash across the chimney-colored basin
Of sky, the plug pulled out and all the light

Gone up the hole. I swig an icy gulp
Of cola-flavored Bacardí. Its brown,
Sweet taste is all that glows. The black and sudden
Moon lasts for seven minutes. Back in town,

Where blackbirds curse the lake along the shore,
Some boys hold up a lifeless fish to blame
The sky for scaring it to death. They say
The water's been bewitched, it's not the same.

I pass a man outside the *pulquería.*
He's stony, like a saint without a shrine,
Propped up against the chipped, celestial wall.
I love the way his shadow blends with mine.

LESLIE MONSOUR

82

FLAMMARION WOODCUT PILGRIM REDUX

✴

He scans the sky and wonders if the Hubble
will burst (or not) the quintessential bubble,
plotting new data on a deep field chart
light years removed from any human heart.

CATHERINE CHANDLER

Flammarion woodcut   The Flammarion woodcut (or engraving) appeared in an 1888
book by French astronomer Nicolas Camille Flammarion. It depicts a man kneeling at a
place where the Earth meets the sky and looking through to a realm of clouds, fires, suns,
and a double wheel (see frontispiece).

83

THE LONELIEST ROAD

✳

Another planet grows and shrinks away,
the heliosphere an ebbing memory,
you streaking like a wayward gamma ray.
Around your vessel blooms a potpourri
of comet, nebula, dark energy
rushing you through the void, accelerating,
all you've ever cared for quickly fading.

What road is lonelier than the universe?
For decades one could sail and never stumble
across another soul. Things could be worse.
Distracted, you could accidentally bumble
too close to a cosmic gullet and wildly tumble,
yet really no more lost than where you coast
past eagle, spider, witch-head, horsehead, ghost.

Though wandering through space entails great risk,
you have no choice—the sun's begun to swell.
While moving at velocities as brisk
as jets of interstellar wind, you smell
the rabbitbrush, the desert breezes, dwell
on sounds of soughing yucca palms and creeks,
glimpse bighorn bounding boulders, rusty streaks

of sunsets. As you near the edge of space,
you think of the stone tools your forebears used
while breathing mayfly lives, a vanished race
in tune with wilderness; and, though you've cruised
for torrents of time now down this road suffused
with radiation, your single mutant eye
still sees, not stars, but fireflies in July.

MARTIN ELSTER

THE MAGICIAN'S BASHFUL DAUGHTER

✳

The moon looks kindly on this slender reed.
Fair and fairylike, and like the moon
When, thin as air, it braves the afternoon,
Advancing while appearing to recede,
The bashful daughter of Le Grand David
Appears onstage to disappear, for soon
She'll step inside the coffin-like cocoon
To be sawn through, and not be seen to bleed.

Reopening the door, her father beams
With more than showman's pride, to find her sound.
Her slippered feet touch lightly on the ground.
She smiles, with braces on. How real she seems.
Oh but the moon is swift to make its round,
And she is only changefulness and dreams.

ALFRED NICOL

85

OLBER'S PARADOX

*

The heavens hold more stars than earth has grains
of sand, and, given time, each tiny sun
combined should make a world where starlight stains
the sky bright white, where dark would be undone.
And yet the night remains. The dim stars gleam
their separate ways, and constellations drawn
connect their dots, while under them we dream
and sleep, then wake to such a thing as dawn.
The universe, expanding since its birth,
is larger, older than its light; sublime,
the force that keeps this constant day from earth—
the same that measures out our years—is time:
the limitation that provides us night
and saves us all from unremitting light.

ROBERT W. CRAWFORD

Olber's Paradox asks why, when the sky is full of stars, is it still dark at night? The first
scientifically reasonable answer to this question was given by American poet Edgar Allan
Poe.

86

JANUARY 22, 2003, OR THE DAY NASA SENT ITS LAST OFFICIAL SIGNAL TO *PIONEER 10*

✳

A budget crunch, but cake for everyone.

Abandonment, but words to soothe the blow:
venerable, plucky, bold.

Deep space and unrequited beeping.

Some said the shape of the probe squeezed from the tip
of an icing gun was nothing short of lovely.
Some said feeble cry and whimper.

Everyone had a slice.

DONNA KANE

Pioneer 10   *Pioneer 10*, an American space probe, was launched in 1972.

87

ZERO GRAVITY: A SPACE REQUIEM

✳

In memory of my sister-in-law Jacqueline Badham (1944–1997)

and to commemorate the voyage of my cousin Joe Tanner and the crew of Space Shuttle STS-82 to repair the Hubble Space Telescope (February 1997)

'The easiest way to think of the universe is as a sphere which is constantly expanding so that everything is getting farther away from everything else.'

SPACE FACTS

I   PROLOGUE

We watched you go
in glory: Shuttle,
comet, sister-in-law.

The one came back.
The other two
went further. Love's an attack

on time. The whole damn thing
explodes, leaving
us with our count-down days

still more than zero.
My theme is change.
My point of view

ecstatic. See how speed
transforms us? Didn't you know
that time's a fiction? We don't need

it for travel. Distance
is a matter of seeing;
faith, a science

of feeling faint objects.
Of course, this is no
consolation as we watch you go

on your dangerous journeys.
This out of mind
hurts badly when you're left behind.

Don't leave us.
We have more to say
before the darkness. Don't go. Stay

a little longer. But you're out of reach
already. Above us the sky
sees with its trillion trillion eyes.

II

Day one at the Wakulla Beach Motel.
There are sixty of us here for the launch.
The kids have found lizards down by the pool,
been shopping in Ron Jon's. I mooch
and admire new body boards. My afternoon's
spent watching them surfing. The sea's my skirt,
breakers my petticoat. Along the seams
of rollers I watch cousins fall without getting hurt
as pelicans button the evening's blouse
into the rollers. The brothers wait,
each head a planet in the shallows' blaze.
They're blind to all others, are a constellation whose gravity makes them surge forward, race

one another, arms flailing, for the broken spume
where they rise, bodies burnished, run back to the place
agreed on for jumping. They laugh in the foam
and I see eclipses, ellipses in the seethe
of a brash outer space. But here they can breathe.

III

It looks like she's drowning
in a linen tide.
They bring babies like cameras
to her bedside

because they can't see dying.
 She looks too well
to be leaving. She listens
to anecdotes we tell –

how we met and got married.
She recounts a story:
her friend went stark mad
carrying, feeding, bleeding – all three

at once. She tried to bury
herself in Barry Island sand.
Her prayer plant has flowered
after seven years. She sends

Robert to fetch it from System St.
She thinks a bee sting
started the cancer.
We can't say a thing.

IV

Bored early morning down on Cocoa Beach,
the kids build castles. I know my history,
so after they've heaped up their Norman keep
(with flags of seaweed) I draw Caerphilly's
concentric fortress. Five-year-old Mary,
who's bringing us shells as they come to hand,
announces, surprised: 'I am the boss of me.'
She has a centre. In our busy sand
we throw up ramparts, a ring of walls
which Sarah crenellates. Being self-contained
can be very stylish — we plan boiling oil!
But soon we're in trouble with what we've designed.
So much for our plans to be fortified.
Our citadel falls to a routine tide.

V

First time I saw the comet, I finally knew
that I'd always love him. I watched it go,

dead starlight headed for a dying sun
then away into darkness. It was gone

before we knew what its brilliance meant,
a human moment in immense

spirals of nothing. I feel his pull
in my blood salts. The comet's tail

is a searchlight from another point,
and the point is once you've given your heart

there are no replacements. Oh, your soul,
if that can escape from its own black hole.

VI

Last suppers, I fancy, are always wide-screen.
I see this one in snapshot: your brothers are rhymes
with you and each other. John has a shiner
from surfing. Already we've started counting time
backwards to zero. The Shuttle processed
out like an idol to its pagan pad.
It stands by its scaffold, being tended and blessed
 by priestly technicians. You refuse to feel sad,
can't wait for your coming wedding with speed
out into weightlessness. We watch you dress
in your orange space suit, a Hindu bride,
with wires like henna for your loveliness.
You carry your helmet like a severed head.
We think of you as already dead.

VII

Her voyage is inwards.
Now looking back
is a matter of passing events.
She makes for the dark

of not being human.
In turn she recalls
giving birth to Robert;
further back, a fall

while pregnant, the bathroom floor
of slate that saved her.
Silk parachute dresses
just after the war.

A bay tree, a garden, Victoria plums.
Then nothing before the age of ten
when a man attacked her.
It must be that pain

accelerates something.
Her speeding mind
leaves us in the present,
a long way behind.

VIII

Thousands arrive when a bird's about to fly,
crowding the causeways. 'Houston. Weather is a go
and counting.' I pray for you as you lie
on your back facing upwards. A placard shows
local, Shuttle and universal time.
Numbers run out. Zero always comes.
'Main engines are gimballed' and I'm
not ready for this, but clouds of steam
billow out sideways and a sudden spark
lifts the rocket on a collective roar
that comes from inside us. With a sonic crack
the spaceship explodes to a flower of fire
on the scaffold's stamen. We sob and swear,
helpless, but we're lifting a sun
with our love's attention, we hear
the Shuttle's death rattle as it overcomes
its own weight with glory, setting car alarms
off in the Keys and then it's gone
out of this time zone, into the calm
of black and we've lost the lemon dawn
your vanishing made. At the viewing site
we pick oranges for your missing light.

IX

The day before
she went she said
'Nothing matters.'
Now that she's dead

she's wiped herself off
our neural screens.
We no longer reach her.
But Jacqueline —

not her body,
nor her history,
nor our view of her —
now she's free

of her rubbish,
explodes on the eye
that perceives faint objects
on an inner sky.

She's our supernova,
sends joyous light
out of her ending.
To our delight

we fell neutrinos
from her ruined core,
can't take our eyes
off her stunning star.

X

Drew trips over his shadow by the pool
but picks himself up. We keep TVs on
like memorial flames, listen as Mission Control
gives cool instructions. You are a sun
we follow, tracking your time over Africa,
a fauvist desert. We see you fall
past pointillist clouds in the Bahamas,
past glaciers, silent hurricanes, the Nile.
We're all provincials when it comes to maps
so we look out for Florida. The world's a road
above you – but you have no 'up',
only an orbit as you dive towards
an opal Pacific, now you see dawn
every ninety minutes. The Shuttle's a cliff
that's shearing, you on it, every way's 'down',
vertiginous plunging. It is yourself
you hold on to, till you lose your grip
on that, even. Then your soul's the ship.

XI

The second time the comet swung by
the knife went deeper. It hissed through the sky

like phosphorus on water. It marked a now,
an only-coming-once, a this-ness we knew

we'd keep forgetting. Its vapour trails
mimicked our voyage along ourselves,

our fire with each other, the endless cold
which surrounds that burning. Don't be fooled

by fireworks. It's no accident that *leave*
fails but still tries to rhyme with *love*.

XII

Only your eyesight can be used in space.
Now you've captured the telescope, nebulae
are birthmarks on your new-born face.
The sun's flare makes a Cyclops eye
on your visor. The new spectrograph
you've installed in the Hubble to replace the old
makes black holes leap closer, allows us to grasp
back in time through distance, to see stars unfold
in nuclear gardens, galaxies like sperm
swirled in water, rashes of young hot stars,
blood-clot catastrophes, febrile swarms
of stinging explosions. But what's far
doesn't stop hurting. Give me a gaze
that sees deep into systems through clouds of debris
to the heart's lone pulsar, let me be amazed
by the red shifts, the sheer luminosity
that plays all around us as we talk on the beach,
thinking there's nothing between us but speech.

XIII

What is her vanishing point?
Now that she's dead
but still close by
we assume she's heard

our conversations.
Out of sight? Out of mind?
On her inward journey
she's travelled beyond

the weight of remembering.
The g-force lifts
from her labouring chest.
Forgetting's a gift

of lightness. She's sped
vast distances
already, she's shed
her many bodies —

cancer, hope, regard,
marriage, forgiving.
Get rid of time
and everything's dancing,

forget straight lines,
all's blown away.
Now's honey from the bees of night,
music from the bees of day.

XIV

There are great advantages to having been dead.
They say that Lazarus never laughed again,
but I doubt it. Your space suit was a shroud
and at night you slept in a catacomb,
posed like a statue. So, having been
out to infinity, you experienced the heat
and roar of re-entry, blood in the veins
then, like a baby, had to find your feet
under you, stagger with weight, learn to cope
again with gravity. Next came the tour
of five states with a stopover in Europe.
You let people touch you, told what you saw.
This counts as a death and a second birth
within one lifetime. This point of view

is radical, its fruit must be mirth
at one's own unimportance and now, although
you're famous, a 'someone', you might want much less.
Your laughter's a longing for weightlessness.

XV

Last sight of the comet. The sky's a screen
riddled with pinpricks, hung in between

me and what happened — a room not quite
hidden from me. Hale-Bopp's light

says something dazzling's taking place beyond,
involving moving. My mind

is silver nitrate, greedy for form
but I fail to grasp it here in this gloom.

Memory's a crude camera.
I wish you were seared on my retina

so I was blind to anything less
than your leaving. But the darkness

is kind. Dawn will heal with colour
my grief for your self-consuming core.

XVI   EPILOGUE

A neighbourhood party
to welcome you home
from the Shuttle's tomb.

It's a wake in sunshine —
kids dive-bomb the pool.
My sense of scale's

exploded. Now I wear
glass beads like planets.
In my ears

are quasars. I have meteorites
for a bracelet, a constellation
necklace so bright

that, despite dark matter
in the heart,
I'm dazzled. 'Here' and 'there'

have flared together. A nonchalant father
throws Saturn rings.
Dive for them now and find everything.

GWYNETH LEWIS

88

3 NIGHTS OF THE PERSEIDS
✳

The night sky's space-debris is radiant
if you can see it, every meteor
a casualty or lost cause in this shower
unsurpassed for 3 nights every August,
witnessed from my balcony: the Perseids.
A comet's castoffs breach the atmosphere,

more shooting stars that graze the stratosphere.
My spotting scope's fixed to a gradient
the better to receive the Perseids
before they vanish, each small light a tear
we might take as a sign or augury
of what's to come, or not: before a shower

of fortune rains or lessens, when its hour
arrives at its own pace. Though not one sphere—
fiery, imperfect—blows past in a gust
of solar wind, small bodies radiate
from somewhere: flying beetles, meteors
of chitin that collide like Perseus,

Andromeda, and fate, pursuing us.
I glimpse a rapid, brief celestial show
in my peripheral vision, meteors
my scope obscured but you saw, sharp and clear,
a stubborn gaze the best ingredient
for awe rewarded, suddenly, in August,

here in Iowa, where you're a guest—
bound for the East Coast when the Perseids
are over, saddened by the radius
our jobs impose, no quick fix in our power...
We live with it. Across the hemisphere,
a waning crescent pales, no meteorites

reported; here, a clockwork cosmos meets
horizons that a greater dark augments.
A jet's lights flash, too far away to fear,
and we can tell its contrails from persistent
trains, gases dissolving in a shower
of ionized traces, white streaks radiant—

How far away they fly: what skies they pierce,
whose lives they meet or tear with every shower
on their august return, still radiant.

NED BALBO

89

CARNAL KNOWLEDGE

✳

Having picked the final datum
From the universe
And fixed it in its column,
Named the causes of infinity,
Performed the calculus
Of the imaginary *i*, it seems

The body aches
To come too,
To the light,
Transmit the grace of gravity,
Express in its own algebra
The symmetries of awe and fear,
The shudder up the spine,
The knowing passing like a cool wind
That leaves the nape hairs leaping.

REBECCA ELSON

90

FERMI'S PARADOX

✳

*The contradiction between the probability of extraterrestrial life and our lack of
contact with it led Enrico Fermi to ask the question, "Where is everybody?"*

We sat together in the dark
and talked about those other worlds
Enrico Fermi thought might be
awash with aliens

because his numbers pointed up
the likelihood of teeming life
among the stars. But where, alas,
did everybody go?

You'd think, perhaps, the prodigies
would come to us or play upon
our signals and reciprocate
—though not if they were bugs.

It could be that they flamed out
the same way we might disappear,
in Malthusian catastrophes
we bring upon ourselves,

or maybe, by design, they've gone
to hide beyond Andromeda
because they realized long ago
how frequently we lie.

We're not worth knowing in the end,
a filthy, biomechanical,
weapons-bearing form of life
that builds amusement parks.

Of all the heavens haven't said,
the best by far I think is this,
that we, together in the dark,
aspire not to care.

JOHN FOY

91

HERE NOW THE SUN: A POEM FOR VALENTINA TERESHKOVA, THE FIRST WOMAN IN SPACE

✳

I. Ready for launch

The suit is working well.
The inflow stream is working well.
I'm ready for launch.
I feel excellent.
Everything is normal.
I'm not a delicate lady.
Everything is normal on board.
I'm ready for launch.
I'm taking up the initial position.
Feeling excellent.

II. Launch

The vehicle's moving smoothly,
vehicle's moving smoothly.
I feel excellent.
Vehicle's moving well.
I feel good.
I feel good.
I see the Earth on the porthole.
I feel excellent.
The Earth is very beautiful.
The vehicle is moving smoothly.
I see the Earth in the porthole,
slightly obscured by clouds.

III. Orbit
I'll do everything that I need to do.
I don't understand.
I didn't see anything.
I feel excellent.
The clock is moving.
I see the horizon through the observation port.
I see the Earth in the observation port.
I feel excellent.
All systems on the vehicle are working perfectly.
Everything is excellent,
I hear you perfectly.

IV. The other cosmonaut
I hear you perfectly,
I feel excellent.
I feel excellent, excellent.
I'm approaching Cape Horn. At the outer ring...
The little star disappeared, wasn't that you?
Don't go far from me, my friend.
I can't see the Moon.
The stars are passing further up.
I am seeing such a bright star.

V. The ships are on their way
The vehicle is responding perfectly, perfectly.
Roger.
Roger.
From the southern point I called him,
he's silent,
from the north,
the same.
At our harbour the ships are silently smoking...

Can you hear?
For the real boys, the harbour is the native home,
comrade to comrade,
they'll always stand together.
And far far away,
the ships are on their way,
and all who are young at heart,
stand shoulder to shoulder.

VI. Fourth orbit
19 hours 25 minutes.
I sang songs for him.
In the centre,
such a blue spot.
Here now the Sun,
so orange, not red,
not light red, but
orange.
I'm also feeling excellent.
Here now the Sun,
visible and lit up.
In the outer ring
the horizon is visible.
It's a very beautiful sight.
At first it's light blue,
then lighter,
then dark...

VII. Greetings to all the women of the world
Soviet women!
Greetings to all Soviet women.
I wish you personal good luck
and great success.

Women of the world!
Greetings to you from space.
I wish you good luck
and success...

VIII. The flight is normal
Cabin pressure 1.15
Humidity 61 percent
Temperature 23 degrees
Carbon dioxide 0.1
Oxygen 250
Pulse 84-90-100
Breathing 22
I feel excellent.
See you soon in the homeland!
I hear you perfectly, perfectly.
The flight is proceeding normally.
All systems of the ship are working perfectly.
I feel excellent.
I hear you.
I'm waiting.
Everything is excellent.
The spaceship is working perfectly.
I'm in good spirits.
I feel excellent.
I hear everything well.
The flight is normal.
All systems on the ship are working perfectly.
Pressure in the suit 1 atmosphere
Humidity 40 percent
Temperature 28 degrees

Carbon dioxide 0.2 percent
Oxygen 200
All systems on the ships are working excellently.
I feel excellent.

IX. Dear Nikita Sergeyevich

Dear Nikita Sergeyevich!
I will use all my strength and knowledge to fully complete the flight.
'Till we meet again soon on our Soviet land.
Moscow, Kremlin.
I am reporting.
Dear Nikita Sergeyevich Khrushchev.
The flight is proceeding normally.
All systems on the ship are working perfectly.
I feel excellent.
Thanks to all the Soviet people.
See you soon in the homeland!
Dear Nikita Sergeyevich,
deeply touched by your attention.
With all my heart.
Dear Nikita Sergeyevich!
I will use all my strength and knowledge to fully complete the flight,
'Till we meet again soon on our Soviet land.

X. Shadows

There aren't enough fingers to block the Sun.
It's very sunny, difficult to see,
at the present a very bright sun,
illuminating the very high clouds...
the horizon above the bright clouds
transitions into shadows.
The dark sky is visible in the survey viewport.
The flight is proceeding normally.
I feel excellent.

XI. This is Chayka

This is Chayka. Over.
This is Chayka. Over.
This is Chayka. Over.
This is Chayka. Over.
This is Chayka. Over.
This is Chayka. Over.
This is Chayka. Over.
This is Chayka. Over.
This is Chayka. Over.

ALICE GORMAN

92

SPACE JOURNAL: SERENDIPITY

✳

En route to Andromeda
I launch luminous spheres
into the black desert of space
to measure the expansion
of the Universe

Back to the garden
The air sparkles
Hildegard Behrens singing
Brünnhilde's dilemma

Resist the urge to touch
rainbow fish in the lily pond
Climb the aspen tree
Its eyes latch on to my toes

Light takes two million years
to reach Andromeda
The Universe is almost empty

A cat's reflection gazes back at me
from the dome's translucent dark vault

YUN WANG

93

SERGEI KRIKALEV ON THE SPACE STATION MIR

✳

> *this is for those people*
> *that hover and hover*
> *and die in the ether peripheries*
>
> Michael Ondaatje, "White Dwarfs"

My name is Sergei and
my body is a balloon.
I want to come down. I
tie myself to things.

My eyes try to describe your
face, they have forgotten.
My ears echo your voice.

I am a star, you can
see me skating on
the dome of night. My blades
catch sun from
the other side of earth.

Days last an hour and a half.
No one else lives here.
My country has disappeared,
I do not know where home is.

I am a painter standing back.
I watch clouds heave like cream
spilled in tea, I see
the burning parrot feathers
of the Amazon forests,
ranges of mountains are
scales along the hide
of the planet, the oceans
are my only sky.

This is my refuge. There is
no one else near me.
Do you understand what that means?

Elena, I am
cold up here.
I hang over Moscow and
imagine you in our flat
feeding little Olga
in a messy chair.
When I drift out of signal range
I do things you
don't want to hear about.

These feet do not know
my weight. A slow
balloon bounces off the walls.
I do not feel like I am flying.

I want to come back and
swim in your hair.
I want to smell you.
I want to arrive in the world
and know my place.
Think of me. I am yours adrift.

Let me describe
my universe. I can see for years.

JAY RUZESKY

Krikalev, a Soviet cosmonaut, was stranded on board the space station Mir in 1991 when
the Soviet Union collapsed. His return was delayed, and he stayed in space twice as long as
originally planned, for 311 days.

94

MY FATHER'S KITES

✳

were crude assemblages of paper sacks and twine,
amalgams of pilfered string and whittled sticks,
twigs pulled straight from his garden, dry patch

of stony land before our house only he
could tend into beauty, thorny roses goaded
into color. How did he make those makeshift

diamonds rise, grab ahold of the wind to sail
into sky like nothing in our neighborhood
of dented cars and stolid brick houses could?

It wasn't through faith or belief in otherworldly
grace, but rather a metaphor from moving
on a street where cars rusted up on blocks,

monstrously immobile, and planes, bound
for that world we could not see, roared
above our heads, our houses pawns

in a bigger flight path. How tricky the launch
into air, the wait for the right eddy to lift
our homemade contraption into the sullen

blue sky above us, our eyes stinging
with the glut of the sun. And the sad tangle
after flight, collapse of grocery bags

and broken branches, snaggle of string
I still cannot unfurl. Father, you left me
with this unsated need to find the most

delicately useful of breezes, to send
myself into the untenable, balance my weight
as if on paper wings, a flutter then fall,

a stutter back to earth, an elastic sense
of being and becoming forged in our front
yard, your hand over mine over balled string.

ALLISON JOSEPH

95

C. E.

✳

(To Keats, in the Empyrean)

Conquistador and watcher of the skies
in one—*Neil Armstrong!* you might well exclaim,
and contemplate retiring Cortez' name.
It's always tempting to contemporize,
and God knows you have time, now, to revise,
but exploration doesn't have the same
cachet it did in your day. No, we came,
we saw, we trudged home, tired, to rest our eyes.
If you were here to speak out loud and bold
and reinvigorate us with a rhyme...
Keats! We are more sub-lunar, less sublime,
than ever, now; the heavens leave us cold.
And presently *the heavens shall be rolled*
together as a scroll. We're marking time.

✳

Re the above, a friend claims I misgauge
the Zeitgeist—says that, to his way of seeing,
we're living in another Golden Age
of exploration, the chief difference being

that missions by and large, now, are unmanned
or telescopic. Do I really mean,
he wants to know, that a new world's less grand
just because we can't be there on the scene,

yet, to plant flags? And Keats, he emphasizes,
praises as much the watchers of the skies as
the *veni, vidi vici*-s of this world.

Fair enough. But "unmanned"...its connotation
Of weakness, cowardice, emasculation...
Olympus Mons: The Stars and Stripes unfurled!

*

(To Auden, in Heaven)

Well, you're too much: "It's natural the boys
should whoop it up for / so huge a phallic triumph"—
so much for that first footstep on the Moon!
God, but that late, avuncular tone cloys!
No offense, Wystan, but when I lose *my* oomph,
I hope I'll have the tact not to impugn
the motives of our brightest minds like *that*.
"I once rode through a desert / and was not charmed..."
Fine, and most people probably share your taste:
but wasn't every human habitat—
including the small plot that Adam farmed
just east of Eden an unpeopled waste?
And don't you think there's something to be said
for finding a new world already dead?

*

Paradise Lost X, 649–662

While the Creator calling forth by name
His mightie Angels gave them several charge,
As sorted best with present things. The Sun
Had first his precept so to move, so shine,
As might affect the Earth with cold and heat

Scarce tollerable, and from the North to call
Decrepit Winter, from the South to bring
 Solstitial summers heat. To the blanc Moone
Her office they prescrib'd, to th' other five
Thir planetarie motions and aspects
In Sextile, Square, and Trine, and Opposite,
Of noxious efficacie, and when to joyne
In Synod unbenigne, and taught the fixt
Thir influence malignant when to showre...

✻

They literally exist, Adam and Eve.
Not only that, it turns out they're *Caucasian,*
or so you might be tempted to believe,

judging from the engraved gold invitation
we sent with Pioneers 10 and 11
four decades back. Asking for an invasion,

If you ask me. (I mean, good God, there's even
a crude map showing the earth's location...)
Why, given human history, believe in

the good intentions of a civilization
no one's seen hide nor hair of, much less met?
Nanking, Dachau, the Trail of Tears, Tibet—

I could go on. As far as annihilation
goes, it may be we ain't seen nothin' yet...

✻

Even for Abraham in Ur before
the days of light pollution, there were more
grains of sand congregated on the shore
than there were stars the naked eye could see;

Still, why split hairs? From where the old man stood
either sky or strand looked like a good
likeness of the well-nigh infinitude
God promised him for his posterity.

For his part, God could well afford to be
generous, within limits. Strand and star
would after all not be the things they are

if they were not defined, respectively,
by waves that immemorially erase
our footsteps, and by airless, unkind space.

*

"A cloud received him," Luke says, matter-of-factly.
The sort of thing von Däniken et al.
would cite as proof that aliens (our "gods")
once walked the Earth (and water, for that matter).

Well, what do you think happened then, exactly?
Were his disciples just hysterical?
(Massively so?) Or glory-seeking frauds?
Did a redactor add this passage later?

He appeared to two disciples en route to Emmaus
and there, when he had broken bread with them,
vanished—simply, tastefully—from sight.

Why then resort to something as outré as
this one-manned lift-off from Jerusalem.
Heaven? There: Look: Second star to the right...

*

(For A.D. Hope, in the Elysian Fields)

Hoping if still from deserts prophets come...
A. D. Hope, from his otherwise grim poem
"Australia." Who knows? Early colonists
on Mars may want the same from their new home,

the key distinction being that in their cases
the whole world outside the airlocked oases
they live in will require renovation
before the mildest of its desert places

is mild enough for one to wander there
in anything like sandals and camel hair,
or from the resurrected rivers sinners
to stand up shining, gasping at thin air—

if in the third millennium C. E.
there's any place at all for prophecy.

＊

Our Quasar television wore a layer
of crackling, dry, slightly resistant air
extending out a half inch from its face.

Seated (my mother always said) too near,
I'd hold one palm against that atmosphere
and make believe it was the edge of space.

We were becoming a space-faring race.
Capsule suspended from its parachute—
How hard would it splash down? And would it float?—
breath held till all three astronauts were out...

And when the set was switched off for the night,
the picture fell back at the speed of light
into a point of incandescent white,
then winked out: a dark mirror took its place.

*

(To Neil Armstrong, *en route*)

Into the poem, a comet with a head
of black ice, trailing newsprint for a tail,
comes the report that Neil Armstrong is dead.

What should the hero's funeral entail?
Would we could order an eclipse at noon
or build a funeral ship with solar sail

or even send a man back to the moon
to lower to half-mast the flag he planted…
The undiscovered country from which none

returns (though there are rumors some have, granted)
receives this quiet hero quietly,
with the small, private service that he wanted.

He leaves a plaque, a flag, some stray debris,
and his light footsteps on Tranquility.

*

It was the quiet that made the stars seem nearer,
that and the fact the air was so much clearer
and that we were at a higher elevation
and far enough away from civilization
we'd gotten out from underneath the dome
of light it raised above itself back home.
Nearer: meaning brighter, meaning so

many stars it could give one vertigo;
quiet: though maybe quiet isn't quite
right for the electronica of night,
the crickets trilling droid-like in their black
body-armor in the grass out back
while the star-pictures, in response to them,
descended like a new Jerusalem.

*

Sunsets of dusty, gold light through the clouds,
breaking in cataract on cataract
of sheer, transfiguring, translucent shrouds.
God's country, our mother called it, and, in fact,

all that was missing were the angels, saints
and conquering Christ depicted in the dome
above the altar of our church back home.

What self-respecting artist these days paints
the immaterial realities
in images as down to earth as these?

Or for that matter would have risked the cluster
of weathered, wordless tombstones by the way,
shining like tablet moons with a late luster
at just that time of year, that time of day?

*

I told my sister each star in the night
was a far sun, some of them probably
with planets like our own, too small to see
even with telescopes, planets that might

have people on them and that, if there were,
it might be that on one of these another
little girl on a hillside with her brother
and father stood there looking back at *her*.

It was a lot to take in. Her eyes grew
round and bright, then, pointing up, she said
I know who lives there, and, when we asked who,

said, *Jesus*.
        I will climb to that same spot
to watch the resurrection of the dead
and greet Christ, the returning astronaut.

BILL COYLE

ON THE NEAREST PASS OF MARS IN 60,000 YEARS

✳

War or Strife—yes, you were always painted
Incarnadine, hematic, flushed with passion,
Sanguine—we depicted you acquainted
With ruby hues the rage in mortal fashion.
And yet to see you ever closer, rolling
Elliptical through emptiness, our gazes
Are met now with a gaze past our controlling,
Red as an eyeball through which blood amazes,
And stony blind. Although we have created
Gods and goddesses of loathing, doting,
They neither love nor hate us, are defeated
By telescopes that taper into nothing,
A stare reflecting on itself, a pleasure
Cold and ferric, nothing we can treasure.

A. E. STALLINGS

DR. RENDEZVOUS TAKES COMMUNION ON
THE MOON

✳

—Buzz Aldrin, Apollo 11, July 1969

Later, the fire of re-entry. Later, depression
medicated by drinking. Later, sobriety, therapy,
the final wrestling free from the prison
of a hard father's surveilling regard. Later still,
past seventy, a left hook to lay flat out
a moon hoax conspiracy theorist
who called him a liar, a coward.

✳

Instead of test-pilot school, he chose
the doctoral dissertation on rendezvous techniques
for manned orbiting vehicles, earning the nickname—
one part admiration, one part
the cocksure bully's jab at any egghead.

But when he solved the unsolvable problem
of Gemini's required spacewalk—not with brute force
or test pilot bravado, but with a scientist's cold,
slow methodology—he found a place
in that uneasy brotherhood.

✳

After the landing, before the one small step,
he pulled from his pocket the tiny chalice,
the vial, the slender wafer—spoke reverent words
about *the vine* and *the fruit*,
drank and ate the blessed sacrament.

Where most others saw—could only see—
two hopelessly separate and reeling vessels,
he knew enough of both math and mystery,
of both faith and reason, to work out
a slender, algorithmic prayer
to guide them into alignment,
to devise the rendezvous required.

LIZ AHL

98

✳

5.

When my father worked on the Hubble Telescope, he said
They operated like surgeons: scrubbed and sheathed
In papery green, the room a clean cold, a bright white.

He'd read Larry Niven at home, and drink scotch on the rocks,
His eyes exhausted and pink. These were the Reagan years,
When we lived with our finger on The Button and struggled

To view our enemies as children. My father spent whole seasons
Bowing before the oracle-eye, hungry for what it would find.
His face lit-up whenever anyone asked, and his arms would rise

As if he were weightless, perfectly at ease in the never-ending
Night of space. On the ground, we tied postcards to balloons
For peace. Prince Charles married Lady Di. Rock Hudson died.

We learned new words for things. The decade changed.

The first few pictures came back blurred, and I felt ashamed
For all the cheerful engineers, my father and his tribe. The second time,
The optics jibed. We saw to the edge of all there is—

So brutal and alive it seemed to comprehend us back.

TRACY K. SMITH

99

RELATIVITY

✳

for Stephen Hawking

When we wake up brushed by panic in the dark
our pupils grope for the shape of things we know.

Photons loosed from slits like greyhounds at the track
reveal light's doubleness in their cast shadows

that stripe a dimmed lab's wall—particles no more—
and with a wave bid all certainties goodbye.

For what's sure in a universe that dopplers
away like a siren's midnight cry? They say

a flash seen from on and off a hurtling train
will explain why time dilates like a perfect

afternoon; predicts black holes where parallel lines
will meet, whose stark horizon even starlight,

bent in its tracks, can't resist. If we can think
this far, might not our eyes adjust to the dark?

SARAH HOWE

100

INCLUDED BY NASA IN A JULY 2021 MISSION
TO JUPITER

✳

  I'm writing to you from a world you'll have
a hard time imagining, to a world I can't
picture no matter how hard I try. Do you still
have birds that wake you up in the morning
with their singing and lovers who gaze at the
stars trying to read in them the fate of their love?
If you do, we'll recognize one another.

CHARLES SIMIC

Acknowledgments

✳

Ned Balbo: "3 Nights of the Perseids" was first published in *3 Nights of the Perseids* (University of Evansville Press, 2019). Reproduced by permission of the author.

Catherine Chandler: "Flammarion Woodcut Pilgrim Redux" from *Lines of Flight* © Catherine Chandler, 2011. Used by permission of Able Muse Press.

Bill Coyle: "C. E." was originally published as only the last two sections in *Able Muse*. Reproduced by permission of the author.

Robert W. Crawford: "Olber's Paradox" from *Too Much Explanation Can Ruin a Man* (David Robert Books, 2005). Reproduced by permission of the author.

Dick Davis: Poem "I Saw the Green Fields of the Sky" from *Faces of Love: Hafez and the Poets of Shiraz*, translated by Dick Davis, © Mage Publishers 2012, www.mage.com.

Rebecca Elson: "Carnal Knowledge" by Rebecca Elson (*A Responsibility to Awe*, 2018) is reprinted by kind permission of Carcanet Press, Manchester, UK.

Martin Elster: "The Loneliest Road" from *Poems for a Liminal Age* (SPM Publications, 2015). Reproduced by permission of the author.

Rhina P. Espaillat: "For My Great-Great-Grandson the Space Pioneer" from *Where Horizons Go* (New Odyssey Press, 1998). Reproduced by permission of the author.

John Foy: "Fermi's Paradox" from *Night Vision* (St. Augustine's Press, 2017). Reproduced by permission of the author.

Robert Frost: "The Star-Splitter" by Robert Frost from *The Poetry of Robert Frost* edited by Edward Connery Lathem. Copyright © 1923, 1969 by Henry Holt and Company. Copyright © 1951 by Robert Frost. Reprinted by permission of Henry Holt and Company. All rights reserved.

Robert Frost: "The Star-Splitter" from *The Collected Poems* by Robert Frost © 1969 Holt Rinehart and Winston, Inc., published by Vintage Books. Extract reproduced by permission of The Random House Group Ltd.

Alice Gorman: "Here now the Sun: a poem for Valentina Tereshkova, the first woman in space" was written using the technique of erasure from the transcripts of her 1963 spaceflight, edited and published in English by Professor Asif Siddiqi (Fordham University, USA): Asif Siddiqi, "Transcripts Give New Perspective on Vostok-6 Mission: The First Woman in Earth Orbit," *Spaceflight* 51 (2009): 18–57. Reproduced by permission of Alice Gorman.

Alan Humm: "Collected songs from the ballad-mistress." Reproduced by permission of Dr. Humm. The poem used primarily the published translations of Michael V. Fox, John L. Foster, Miriam Lichtheim, John A. Wilson, and J. M. Plumley.

Elizabeth Jennings: "Delay" from *The Collected Poems* by Elizabeth Jennings (Carcanet Press, 2012), reproduced by permission of David Higham Associates.

Donna Kane: "JANUARY 22, 2003 OR THE DAY NASA SENT ITS LAST OFFICIAL SIGNAL TO *PIONEER 10*" from *Orrery* (Harbour Publishing, 2020). Reproduced by permission of the publisher.

X. J. Kennedy: "Ladder to the Moon." By permission of X. J. Kennedy.

Len Krisak: "*Carmina 66*" was first published in *Carmina* (Carcanet Press, 2014). Reproduced by permission of the author.

Acknowledgments

Gwyneth Lewis: "Zero Gravity" from Gwyneth Lewis, *Chaotic Angels, Poems in English* (Bloodaxe Books, 2005). Reproduced with permission of Bloodaxe Books. www.bloodaxe.com.

Leslie Monsour: "Rainy Eclipse" from *The Alarming Beauty of the Sky* (Red Hen Press, 2006). Reproduced by permission of the author.

Edwin Morgan: "The First Men on Mercury" from *From Glasgow to Saturn* (Carcanet Press, 1973), also published in *Collected Poems* (Carcanet Press, 1990). Reprinted by kind permission of Carcanet Press, Manchester, UK.

Howard Nemerov: "Witnessing the Launch of the Shuttle Atlantis." By permission of the Howard Nemerov estate.

Alfred Nicol: "The Magician's Bashful Daughter" from *Winter Light* (University of Evansville Press, 2004). Reproduced by permission of the author.

Linda Pastan: "Eclipse" © 2015 by Linda Pastan. Used by permission of Linda Pastan in care of the Jean V. Naggar Literary Agency, Inc. (permissions@jvnla.com)

Linda Pastan: "Eclipse," from INSOMNIA: POEMS by Linda Pastan. Copyright © 2015 by Linda Pastan. Used by permission of W. W. Norton & Company, Inc.

Jay Ruzesky: "Sergei Krikalev on the Space Station Mir" was originally published in *Painting the Yellow House Blue* (Toronto: House of Anansi Press, 1994). Reproduced by permission of Jay Ruzesky.

Tracy K. Smith: Credit: Tracy K. Smith, Part 5 from "My God, It's Full of Stars" from *Such Color: New and Selected Poems*. Copyright © 2011 by Tracy K. Smith. Reprinted with the permission of The Permissions Company, LLC on behalf of Graywolf Press, Minneapolis, Minnesota, www.graywolfpress.org.

A. E. Stallings: "On the Nearest Pass of Mars in 60,000 Years." Copyright © 2006 by A. E. Stallings. Published 2006 by Northwestern University Press. All rights reserved.

Willard Trask: *Classic Black African Poems*; translation copyright Willard Trask/Eakins Press Foundation; used with permission.

Yun Wang: "Space Journal: Serendipity" was originally published by *Abstract Magazine TV*. Reproduced by permission of the author.

Deborah Warren: "The Ballet of the Eight-Week Kittens" from *Dream with Flowers and Bowl of Fruit* (University of Evansville Press, 2008). Reproduced by permission of the author.

Richard Wilbur: "In the Field." By permission of the Richard Wilbur estate.

All poems not credited are in the public domain.

I would like to thank the following for contributing poems and translations to this volume: Liz Ahl, Ned Balbo, Alice Berlingett (all efforts were made to find the estate of Mrs. Berlingett, and I am very grateful to include her poem here), Catherine Chandler, Bill Coyle, Robert W. Crawford, John Curl, Dick Davis, Eakins Press for kind permission to use Willard Trask's translation, Martin Elster, Rhina Espaillat, Michael Ferber, John Foy, Dr. Alice Gorman, Harbour Publishing, Sarah Howe, Dr. Alan Humm, Allison Joseph, A. M. Juster, Donna Kane, X. J. Kennedy, Janet Kenny, Len Krisak, Mage Publishers, Leslie Monsour, Victoria Moul, the Jean V. Naggar Literary Agency, the Howard Nemerov Estate, Alfred Nicol, W.W. Norton, Linda Pastan, Jay Ruzesky, Charles Simic, Yun Wang, Deborah Warren, the Richard Wilbur Estate, and Anton Yakovlev. Without their generosity this book would not exist.

I'd like to thank Bethany Thomas, editor at Cambridge University Press, for believing in this project from the very beginning, and working relentlessly to find a way to make it a reality. I'd also like to thank the University of Cambridge Institute of Astronomy for making it possible to include the poem by Rebecca Elson. George Paul Laver at Cambridge University Press answered endless questions from me about putting together a manuscript. Rhina Espaillat and Deborah Warren not only contributed poems and translations, they translated them on demand. Hannah Campeanu and Robert W. Crawford gave valuable advice at any hour of the day or night throughout the process.

Many thanks to Laurence Marschall, Professor of Physics, Emeritus, at Gettysburg College, for sharing his favorite astronomy poems, a number of which were included in this anthology.

Charles Simic, Professor Emeritus of English at the University of New Hampshire, says he is in daily contact with *Lucy* (the NASA space probe) and that she is bored. Deep thanks to both of them for the wonderful final poem of the book.

Most of all, I'd like to thank Michael Ferber, Professor Emeritus of English and Humanities at the University of New Hampshire. It was his *Romanticism: 100 Poems* anthology for this same series that first gave me the idea for the book. He provided connections, suggestions, edits, answers, and even a translation for the volume. Without his support all throughout the process, this book would not have been possible.

Biographies

✳

Liz Ahl (1970) is the author of *Beating the Bounds* and several other poetry collections. She was born nine months after the first moon landing and has been exploring the history and science of the US space program of the 1960s since seeing *Apollo 13*. She lives in New Hampshire.

Louisa May Alcott (1832–1888) was an American novelist, short story writer, and poet.

Dante Alighieri (1265–1321) was an Italian poet, writer, and statesman.

Matthew Arnold (1822–1888) was an English poet, literary and social critic, and inspector of schools.

Ned Balbo (1959) is the author of *The Cylburn Touch-Me-Nots* (New Criterion Poetry Prize), *3 Nights of the Perseids* (Richard Wilbur Award), *Upcycling Paumanok*, *Lives of the Sleepers* (Ernest Sandeen Prize), *Galileo's Banquet* (Towson University Prize), and *The Trials of Edgar Poe and Other Poems* (Poets' Prize and the Donald Justice Prize).

Anna Laetitia Barbauld (1743–1825) was an English poet, writer, and teacher.

Charles Baudelaire (1821–1867) was a French poet, translator, and critic.

Alice Berlingett (*c.* 1868–unknown) was an American poet who lived in Norfolk, Virginia. Her poems appeared several times in *Popular Astronomy* and *The English Journal*.

Bhāsa (*c.* 300) was a Sanskrit dramatist.

William Blake (1757–1827) was an English poet, painter, engraver, and one of the poets of English Romanticism.

Hjalmar Boyesen (1848–1895) was a Norwegian-American author and college professor.

Tycho Brahe (1546–1601) was a Danish astronomer.

Elizabeth Barrett Browning (1806–1861) was an English poet.

William Cullen Bryant (1794–1878) was an American poet, journalist, and newspaper editor.

Rosalía de Castro (1837–1885) was a Galician poet who wrote in Galician and Spanish.

Catullus (*c.* 84–*c.* 54 BCE), full name Gaius Valerius Catullus, was a Roman poet.

Catherine Chandler (1950) is the author of *Lines of Flight*, shortlisted for the Poets' Prize, and *The Frangible Hour*, winner of the 2016 Richard Wilbur Award. Her sixth collection, *Annals of the Dear Unknown*, a historical verse-tale set in eighteenth-century Pennsylvania, is forthcoming from Kelsay Books.

Geoffrey Chaucer (*c.* 1340–1400) was an English poet and author.

Lady Mary Chudleigh (1656–1710) was an English poet and essayist.

Samuel Coleridge (1772–1834) was an English poet, literary critic, philosopher, theologian, and one of the founders of English Romanticism.

Bill Coyle (1968) won the 2006 New Criterion Poetry Prize for *The God of This World to His Prophet*. In 2010 he received a translation grant from the National Endowment for the Arts. A collection of his translations of the Swedish poet Håkan Sandell, *Dog Star Notations: Selected Poems 1999–2016* (Carcanet Press), was published in 2016.

Robert W. Crawford (1958) is the author of *The Empty Chair* (University of Evansville), winner of the 2011 Richard Wilbur Award, and *Too Much Explanation Can Ruin a Man* (David Robert Books). He has twice won the Howard Nemerov Sonnet Award. He is the Director of Frost Farm Poetry in Derry, NH, which includes the Hyla Brook Reading Series, the Frost Farm Poetry Conference, and the Frost Farm Poetry Prize.

John Curl (1940) is the author of *Ancient American Poets*, his translations and biographies of classical Maya, Aztec, and Inca poets. He studied spoken Maya in Quintana Roo, spoken Nahuatl in Veracruz, and spoken Quechua in Cusco. He has a degree in Comparative Literature from the City College of New York, and resides in Berkeley, CA.

David (*c.* 1042–*c.* 970 BCE) was the biblical king of ancient Israel and Judah.

Dick Davis (1945), Emeritus Professor of Persian at Ohio State University, has written scholarly works on both English and Persian literature, as well as numerous books of translations from Persian and nine volumes of his own poetry. The *Times Literary Supplement* has referred to him as "our finest translator from Persian."

John Donne (1572–1631) was an English poet, writer, politician, and Anglican cleric.

Du Fu (712–770) was a Chinese poet.

Paul Laurence Dunbar (1872–1906) was an American poet, novelist, and writer.

Rebecca Elson (1960–1999) was a Canadian-born astronomer and poet.

Martin Elster (1954), who never misses a beat, was for many years a percussionist with the Hartford Symphony Orchestra. His poems have appeared in numerous literary journals and anthologies in the USA and abroad. A full-length collection, *Celestial Euphony* (Plum White Press), was published in 2019.

Ralph Waldo Emerson (1803–1882) was an American essayist, poet, philosopher, lecturer, and leading Transcendentalist.

Rhina P. Espaillat (1932), a Dominican-born American poet, has published thirteen books, five chapbooks, and two CDs, comprising poetry, essays, and short stories in both English and her native Spanish, as well as translations into and from both languages and several others. Her work appears in numerous anthologies and magazines and has earned many national and international awards.

Michael Ferber (1944) is Professor Emeritus of English and Humanities at the University of New Hampshire. He is the editor of another volume in this series: *Romanticism: 100 Poems.*

John Foy (1960) won the Donald Justice Poetry Prize for his third book, *No One Leaves the World Unhurt* (Autumn House Press). His second book, *Night Vision* (St. Augustine Press), won the New Criterion Poetry Prize. His work has appeared widely in journals and online. He lives in New York.

Robert Frost (1874–1963) was an American poet who was awarded a Congressional Gold Medal and received the Pulitzer Prize four times, in 1924, 1931, 1937, and 1943.

Johann Wolfgang von Goethe (1749–1832) was a German poet, playwright, and novelist.

Alice Gorman (1964) is an internationally recognized leader in the field of space archaeology and the author of the award-winning book *Dr Space Junk vs the Universe: Archaeology and the Future* (MIT Press). Asteroid 551014 Gorman is named after her in recognition of her work in establishing space archaeology.

Hafez (*c.* 1315–*c.* 1390), whose full name was Khwāja Šams ud-Dīn Muḥammad Ḥāfez-e Šīrāz, was a Persian lyric poet from Iran.

Edmund Halley (1656–1742) was an English astronomer and mathematician, known also for his role in the publication of Isaac Newton's *Philosophiae Naturalis Principia Mathematica*. Halley's Comet is named after him.

Thomas Hardy (1840–1928) was an English novelist and poet.

Frances Ellen Watkins Harper (1825–1911) was an American poet, writer, journalist, activist, and lecturer.

Oliver Herford (1863–1935) was a British-born American writer, poet, illustrator, and artist.

Homer (*c.* 750 BCE) was an ancient Greek poet and author.

Laurence Hope (Adela Florence Cory Nicolson) (1865–1904) was an English poet, later known as Violet Nicolson.

Gerard Manley Hopkins (1844–1889) was an English poet and Jesuit priest.

A. E. Housman (1859–1936) was an English scholar and poet.

Sarah Howe (1983) is a Hong Kong-born British poet, academic, and editor. Her first collection of poems is *Loop of Jade* (Chatto & Windus).

Alan Humm (1953) received his doctorate in Religious Studies from the University of Pennsylvania and has taught at various colleges, including LaSalle University (Philadelphia) and Albright College (Reading, PA). His publications include *The Psychology of Prophecy in Early Christianity* (Gorgias Press) and contributions to *The Ashgate Encyclopedia of Literary and Cinematic Monsters* (Ashgate Publishing).

Helen Hunt Jackson (1830–1885) was an American poet, writer, and activist.

Elizabeth Jennings (1926–2001) was an English poet.

Ben Jonson (1572–1637) was an English poet, playwright, and literary critic.

Allison Joseph (1967) is a British-born American poet, professor at Southern Illinois University, and editor.

A. M. Juster (1956) has written poems and translations that have appeared in *Poetry*, *The Paris Review*, *The Hudson Review*, and other journals. He is the only three-time winner of the Howard Nemerov Sonnet Award and has won the Barnstone Translation and Richard Wilbur Awards. His translation of Petrarch's *Canzoniere* (W.W. Norton) will be released in 2023.

Donna Kane (1959) is a writer living in Rolla, BC, Canada. Her poems, short fiction, reviews, and essays have been published widely. Her most recent book of poetry, *Orrery*, was a finalist for the 2020 Governor General's Literary Award.

John Keats (1795–1821) was an English poet and one of the poets of English Romanticism.

X. J. Kennedy (1929) is an American poet, translator, editor, anthologist, and author of textbooks and children's literature.

Janet Kenny (1936) was born in New Zealand and sang professionally in Britain. She co-published *Beyond Chernobyl: Women Respond*, and her poetry collections are *This Way to the Exit* (White Violet Press) and

Whistling in the Dark (Kelsay Books). Married to her late husband for sixty-six years, she lives by the sea in Queensland.

Len Krisak (1948), a poet and classics translator, has just published a volume of his own verse, *Say What You Will*, and his version of the complete *Aeneid*. He lives near Boston.

Charles Godfrey Leland (1824–1903) was an American folklorist, poet, and writer.

Gwyneth Lewis (1959) is a Welsh poet and was the inaugural National Poet of Wales in 2005.

Henry Wadsworth Longfellow (1807–1882) was an American poet and the first American translator of Dante.

Federico García Lorca (1898–1936) was a Spanish poet and playwright.

Lucretius (*c.* 94–*c.* 49 BCE), full name Titus Lucretius Carus, was a Roman poet and philosopher.

John Gillespie Magee, Jr. (1922–1941) was an Anglo-American pilot with the Royal Canadian Air Force in World War II, who was killed during a training flight in England.

Christopher Marlowe (1564–1593) was an English poet and playwright. The passage included comes from version B of *The Tragical History of Dr. Faustus*.

Edgar Lee Masters (1868–1950) was an American poet, attorney, playwright, and biographer.

Claude McKay (1889–1948) was a Jamaican-born American poet, novelist, and journalist, and one of the poets of the Harlem Renaissance.

Michelangelo (1475–1564), full name Michelangelo di Lodovico Buonarroti Simoni, was an Italian sculptor, painter, architect, and poet.

Edna St. Vincent Millay (1892–1950) was an American poet, playwright, and recipient of the 1923 Pulitzer Prize for Poetry.

John Milton (1608–1674) was an English poet, writer, and historian.

Leslie Monsour (1948), an American poet, was born in Los Angeles and raised in Mexico City. The recipient of a fellowship from the National Endowment for the Arts, as well as five Pushcart Prize nominations, Monsour is the author of two poetry collections and the recent *Colosseum Critical Introduction to Rhina P. Espaillat* (Franciscan University Press).

Edwin Morgan (1920–2010), a Scottish poet, translator, and educator, was the first Scots Makar, or Scottish national poet, and received the Queen's Gold Medal for Poetry.

Victoria Moul (1980) is Reader in Early Modern Latin and English at University College London. She also writes and translates poetry.

Howard Nemerov (1920–1991), an American poet, writer, and critic, served as poetry consultant to the Library of Congress and as the Poet Laureate of the United States. He received the 1978 Pulitzer Prize for Poetry.

Alfred Nicol (1956) is the author of *Animal Psalms* (Able Muse), *Elegy for Everyone* (Prospero's World), and *Winter Light* (University of Evansville). His poems have appeared in *Poetry*, *The New England Review*, *Dark Horse*, *Commonweal*, *The Formalist*, *The Hopkins Review*, *The Best American Poetry 2018*, and other literary journals and anthologies.

Linda Pastan (1932) is an American poet, whose fourteenth book of poems, *Insomnia*, published in 2015, won the Towson State Literature Prize. *A Dog Runs Through* was published in 2018 and *Almost An Elegy* is due in 2022. She has twice been a finalist for The National Book Award.

Edgar Allan Poe (1809–1849) was an American poet, writer, critic, and editor.

Alexander Pushkin (1799–1837) was a Russian poet, playwright, and novelist.

John Rollin Ridge (Yellow Bird) (1827–1867) a member of the Cherokee Nation, was an American poet, novelist, journalist, and newspaper editor.

Jay Ruzesky (1965) is a writer and digital media artist. He is the author of three poetry collections as well as a novel, and *After Antarctica: An Amundsen Pilgrimage*, a book of creative nonfiction. He lives on Vancouver Island in British Columbia.

Sappho (*c.* 600 BCE) was a Greek lyric poet from the island of Lesbos.

William Shakespeare (1564–1616) was an English poet and playwright.

Percy Bysshe Shelley (1792–1822) was an English poet, one of the poets of English Romanticism.

Sir Philip Sidney (1554–1586), was an English courtier, statesman, soldier, poet, and patron of the arts.

Charles Simic (1938), a Serbian American poet, was the Poet Laureate of the United States and received the 1990 Pulitzer Prize for Poetry.

Tracy K. Smith (1972), an American poet, author, and educator, was the Poet Laureate of the United States and received the 2012 Pulitzer Prize for Poetry.

Edmund Spenser (*c.* 1552–1599) was an English poet.

A.E. (Alicia) Stallings (1968) is an American-born poet who lives in Greece. She has published four collections of poetry, most recently *Like* (FSG), a finalist for the Pulitzer Prize, and three volumes of verse translation, most recently *The Battle Between the Frogs and the Mice* (Paul Dry Books). Her *Selected Poems* are forthcoming from FSG and Carcanet.

Frank Pearce Sturm (1879–1942) was an English poet and translator.

John Addington Symonds (1840–1893) was an English poet, literary critic, and cultural historian.

Sara Teasdale (1884–1933), an American poet, was, in 1918, the first recipient of the award that would become the Pulitzer Prize for Poetry.

Alfred, Lord Tennyson (1809–1892) was an English poet and the Poet Laureate during Queen Victoria's reign.

Henry David Thoreau (1817–1862) was an American poet, naturalist, essayist, and leading Transcendentalist.

Willard Trask (1900–1980) was a celebrated translator of numerous languages into English, who won the first National Book Award for translation in 1967, the year after the publication of his *Classic Black African Poems* (Eakins Press), which he selected, edited, and translated.

Virgil (70–19 BCE), full name Publius Vergilius Maro, was a Roman poet.

Bertrand N. O. Walker (Hen-tah) (*c.* 1870–1926), a member of the Wyandotte Nation, was an American poet, teacher, and folklorist.

Yun Wang (1964) has authored *The Book of Mirrors* (White Pine Press), *The Book of Totality* (Salmon Poetry Press), *The Book of Jade* (Story Line Press), and *Dreaming of Fallen Blossoms: Tune Poems of Su Dong-Po* (White Pine Press). She is a cosmologist at California Institute of Technology.

Deborah Warren (1946) is the author of *Connoisseurs of Worms* (Paul Dry Books); *Dream With Flowers and Bowl of Fruit* (University of Evansville), Richard Wilbur Award; *Zero Meridian* (Ivan R. Dee), New Criterion Poetry Prize; *The Size of Happiness* (Waywiser Press, Anthony Hecht Poetry Prize); *Ausonius: The Moselle and Other Poems*, translation (Routledge); and *Strange to Say: Etymology for Serious Entertainment* (Paul Dry Books).

Mercy Otis Warren (1728–1814) was an American poet, playwright, historian, and satirist.

Phyllis Wheatley (*c.* 1753–1784), born in West Africa and sold into slavery, was the first African American and one of the first women in the American colonies to publish a book of poetry.

Walt Whitman (1819–1892) was an American poet, journalist, and essayist.

Richard Wilbur (1921–2017), an American poet and translator, served as Poet Laureate of the United States and received the Pulitzer Prize for Poetry twice, in 1957 and 1989.

Ella Wheeler Wilcox (1850–1919) was an American poet and journalist.

Helen Maria Williams (1759–1827) was an English poet, novelist, writer, abolitionist, and translator.

Sarah Williams (1837–1868) was an English poet and novelist.

William Wordsworth (1770–1850) was an English poet, one of the founders of English Romanticism, and Poet Laureate during Queen Victoria's reign.

Anton Yakovlev (1981) Yakovlev's poems have appeared in *The New Yorker*, *The New Criterion*, *The Hopkins Review*, and elsewhere. His latest chapbook, *Chronos Dines Alone* (SurVision Books), won the James Tate Prize. *The Last Poet of the Village: Selected Poems by Sergei Yesenin* (Sensitive Skin Books) was published in 2019.

William Butler Yeats (1865–1939) was an Irish poet and playwright who received the 1923 Nobel Prize for Literature.